D1370473

ROCKETS & MISSILES
OF WORLD WAR III

ROBERT BERMAN & BILL GUNSTON

ROBERT BERMAN & BILL GUNSTON
ROCKETS
& MISSILES
OF WORLD WAR III

Exeter Books

NEW YORK

Copyright © 1983 Bison Books Corp

First published in USA 1983
by Exeter Books
Distributed by Bookthrift
Exeter is a trademark of Simon & Schuster
Bookthrift is a registered trademark of Simon & Schuster
New York, New York

ALL RIGHTS RESERVED

ISBN 0-671-06004-X

Printed in Hong Kong

Contents

1. MISSILE TECHNOLOGY AND HISTORY

Rheintochter (Rhine daughter, or Rhinemaiden) was one of the first SAMs (surface-to-air missiles). This one was photographed on its launcher in Germany in late 1943.

While the speed and sound of rockets and missiles have fascinated man for a thousand years, his survival today is dependent on those weapons never being used. Rockets and missiles of the earliest times had been used as sources of entertainment, as signaling devices and sometimes as simple weapons, but with indifferent results.

Rockets can trace their history back to China and the production of gunpowder in the 11th century. Eventually the gunpowder was packed into metal cylinders and mounted on arrows. These projectiles, although short in range, had some shock value as a result of their fire and sound effects. Over two hundred years later this technology had its first recorded military application when the Chinese broke the Mongol siege of a city by the surprise use of hundreds of rocket arrows against the invading army.

While rockets became popular for their novelty more than anything else, another 600 years had passed before the rocket came back into vogue in military circles. This time, though, the British were on the receiving end in a battle with the Sultan of Mysore, in a part of what is now India, during the late 18th century. Again the weapons were short-range explosive and shrapnel rockets. What was unique about the rockets' use was the special organization the Sultan of Mysore established to control the rockets in the army. This was perhaps the first dedicated missile unit.

The use of rocket power against the British in India encouraged others who were already intrigued with the rocket's potential. One of the more important was Sir William Congreve. Congreve developed a missile along the same principle as the Chinese rockets, but with a combat range of a mile; it was also considerably larger than the Chinese or Indian weapons.

William (later Sir William) Congreve pioneered the unguided bombardment rockets used in large numbers by Britain on land and sea in the early 19th Century.

Congreve's weapons had their first operational test in the first national conflict of our time – the Napoleonic wars. In an assault off Boulogne in 1806, a British fleet of eighteen ships fired thirty tons of incendiary rockets in less than an hour, doing extensive damage. In 1807 the British fleet struck again. This time the British launched nearly 375 tons of rockets against Copenhagen. The attack caused widespread destruction, and soon these kinds of weapons began to be put into service in the French, Austrian, Russian and Italian armies. Rockets were use again by the British in the War of 1812 against the United States and, as improved versions of Congreve's rocket were introduced, these weapons were used by the United States in its war with Mexico in the mid-1840s.

Despite the widespread used of the rocket, its role as a major weapon was shortlived. Cannon and artillery began to be improved to the point where the assault rocket was no longer essential for the modern army. The mobility, range, explosive power and accuracy of artillery pieces in the latter half of the 19th century would signal the diminishing use of the rocket for war for almost 75 years.

The Transition to Modern Missiles

In that interim period the development of rocketry, aided by both industrial and technical innovations, would lay the foundation for both space exploration and strategic weapon arsenals. These developments, however, would not have taken place if it were not for the imagination and endeavor of a group of men from many different countries, who clearly raised the level of research and development of missiles and rockets to that of the science that it is today. We are, of course, referring to K. E. Tsiolkovsky of Russia, Robert H. Goddard of the United States and Herman Oberth of Germany.

Tsiolkovsky was concerned with the theoretical principles of modern rocketry, propulsion and space travel. He confirmed Newton's basic theory of action and reaction as it applied to a rocket leaving the earth and traveling through the atmosphere to reach space. Further studies by Tsiolkovsky revolved around rocket fuels, rocket construction and missile guidance.

Robert Goddard's role in the development of modern rocketry was pivotal. While also developing a large body of theoretical work, Goddard constructed the first liquid-propellant rocket, launched on 16 March 1926. Goddard worked in relative isolation until World War II, when the military requirements of missile flight absorbed all of his efforts. Goddard's contribution to missile flight was underlined by German World War II missilier Wernher von Braun, when he commented that everything his missile development group worked on was drawn from Goddard's activities.

German interest in rockets had a greater following as well as being better organized. Consequently the Germans entered World War II with the lead in missile technology. Herman Oberth was the leader of the Society for Space Flight, echoing the familiar theme of space travel as the lure for missile development. When Hitler came to power the society was disbanded and its

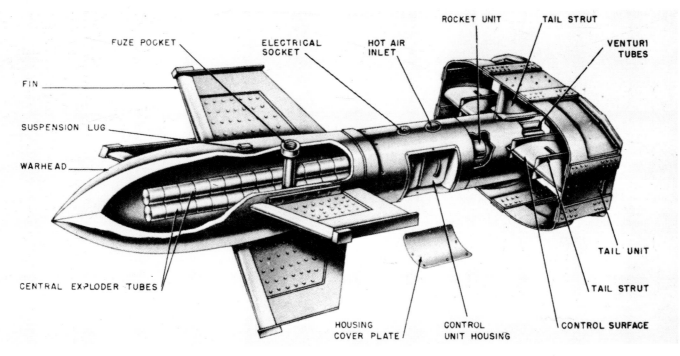

FIN — FUZE POCKET — ELECTRICAL SOCKET — HOT AIR INLET — ROCKET UNIT — TAIL STRUT — VENTURI TUBES — SUSPENSION LUG — WARHEAD — CENTRAL EXPLODER TUBES — HOUSING COVER PLATE — CONTROL UNIT HOUSING — TAIL UNIT — TAIL STRUT — CONTROL SURFACE

Above: X-3 was the world's first guided air-to-air missile, fired from Luftwaffe fighters in 1944 and steered by signals transmitted along thin wires.

Top: Cutaway drawing of another of Germany's impressive guided missiles of World War II, the FX, or FX.1400. This was a 3,461-lb weapon dropped from bombers and steered by radio.

affairs assumed by the army. By 1937 a missile base had been constructed at Peenemünde on the German Baltic coast. Here the chief scientist of the old space society, Wernher von Braun, would develop the V-1 and later the V-2 liquid-propelled rocket.

Missile Technology

The first rockets were basic instruments. They were usually hollow tubes filled with gunpowder and secured to arrows. Nevertheless, these simple devices share some characteristics with a modern Minuteman III ICBM. Both rockets are propelled in the direction opposite to the thrust of the fuel that is being expelled as a gas. This exemplifies Newton's Third Law of Motion:

for every action there is an equal and opposite reaction.

But rockets also required some means of stabilizing their flight. Unlike artillery, where the bores of cannons were beginning to be rifled for improved stability, with improved range and accuracy resulting, old stick missiles relied on a decaying ballistic trajectory based on a host of environmental factors to achieve accuracy.

The first serious approach to the problem of rocket stability was the introduction of vanes in the tails of rockets. When in motion, exhaust gases would pass around the vanes, causing the rocket to spin and inducing stability. Modern missiles would by-pass this method and use gimballed or engine nozzles that pivot to maintain missile balance.

World War II

The development in World War II of the German V-1 cruise missile, the V-2 ballistic missile and the Waterfall surface-to-air missile created the foundation for the space missiles and modern missile weapons developed by the United States and the Soviet Union.

Once the German research effort on space travel became a military effort in 1933, a number of different experimental programs and concepts began to take form. The two major efforts to bear fruit were, of course, the V-1 and the V-2.

While the V-1 and the V-2 had similar military objectives, two different military services, the Air Force and the Army, controlled the development and deployment of the weapons, with neither development group aware of the existence of the other.

Development on all of Germany's missiles took place at Peenemünde on the Baltic coast, in what is now East Germany. The test site was very large and contained all the facilities needed for missile development and testing.

The Air Force V-1 was a small pilotless aircraft propelled by a pulse-jet engine in which the propellant was ignited in spurts. It could carry one ton of explosive some 250 miles at 400 miles per hour, while its altitude varied between 1000 and 7000 feet. Guidance was simple; when the fuel ran out, the V-1 would drop until it exploded on something.

The V-1 concept had been discussed for over twenty years prior to the Air Force's putting the weapon into production through the War Ministry in 1941 and making it operational three years later. In the course of the war some 7300 were launched at Belgium and Britain from the ground and air, but without a guidance system they produced indifferent military results. On one day over 300 V-1s were launched at targets. Importantly, a number of V-1 did not reach their 'target'. Because of a slow ambling flight path, the V-1 could be shot down by

Above: Launch of an A-4 development rocket from Peenemünde test centre in early 1944. These were painted black and white to show up well in ciné film.
Below: Popularly called the V-1, the Fi 103 cruise missile was used in thousands against London and other English targets, with another 2,448 fired against the Belgian city of Antwerp.
Right: An instrumented A-4 (so-called V-2) rocket in its pre-launch gantry at the Peenemünde test centre in 1943.

fighter aircraft and anti-aircraft artillery. Of the 7300 V-1s launched, almost one-fourth were shot down by fighters and one-fifth by anti-aircraft guns. Another five percent were destroyed by disintegration when they ran into barrage balloons. Although they cost less than 10,000 German marks (as compared to 75,000 for the V-2), the V-1 susceptibility to enemy air defense prevented the Air Force from realizing a weapon that could replace its bomber force or have a decisive impact on the war.

The most stunning achievement of Peenemünde was the development of the V-2. From Wernher von Braun's design teams' research would come the first missiles of the Cold War, such as the Soviet SS-3 and SS-6 and the American Jupiter and space rockets: rockets like the one that put the first Sputnik into orbit in 1957 and the Saturn V that took US astronauts to the moon.

In 1933 work began on the various prototype missiles that would eventually become the V-2. The initial test vehicles, such as the Aggregate 1 (A-1), the A-2 tested in 1935 and the A-3 tested in 1937, all had problems. The A-1, for example, had stability and fuel problems while the A-2 and A-3 had crude and inadequate guidance systems. As the war came close, funds from the War Department increased and work progressed accordingly. At the same time the War Department re-emphasized its requirement to strike urban targets in Britain.

Breakthroughs in missile developments were made regularly. Improved mechanical control systems and high-capacity fuel pumps were developed and tested at Peenemünde in expectation of the first V-2 (called A-4 when it was under development) test. The V-2 was tested in the summer and fall of 1942 before the first successful launch on 31 October of that year. Other successes followed, but it would still be two years before the V-2 would become operational. In that time the Germans would test over 1500 V-2s, ensuring reliability once operational.

The V-2 in operational form – like the V-1 – could propel a one-ton charge of high explosive some 250 miles. There the similarity ended, because the V-2, in its ballistic space trajectory, would travel at 3500 miles an hour, making it impossible to defend against.

The initial launch pads for the 12½-ton V-2 were fixed firing platforms at Peenemünde. These launch pads were vulnerable to air attack and during 1943, while the V-2 was still being tested, RAF bombers struck the facility with 2000 tons of bombs, doing considerable damage. As a result production of the missile was moved underground.

One of the more innovative approaches to operating the missiles was conceived of early on in the missile program. Rather than launch missiles from Peenemünde, Army Captain (later General) Dornberger developed a mobile basing concept whereby V-2 batteries, each with three launchers, would set up their missiles after arriving at a pre-selected location in 90 minutes, then fire 15 missiles in the course of a day before moving on to avoid air attack. Thus, Test Battery 444 operated in Poland until the Red Army approached and then

moved to Belgium in preparation for initial operations as Motorized Artillery Section 485. Its first assault, on 5 September 1944, was against Paris. During the rest of 1944 nearly 1600 V-2s were launched, with nearly 500 hitting London and 900 striking Antwerp, where the Allies had a major supply and logistics center. Other targets were against various locations in Europe. The launchings were not without difficulties; guidance and engine problems limited the successful firing rate to about 90 percent. By war's end, with production peaking at around 1000 per month, some 5500 V-2s had been launched: against London (1100), Antwerp (1600) and other objectives from the mobile V-2 rocket army. While not changing the war's outcome, the introduction of the V-2 one year earlier would have complicated Allied invasion plans and could have delayed the war's end even longer. The new era of warfare ushered in by the V-2 produced much data for the US and the Soviet Union in later years. In addition to mobile V-2 units, plans were also developed for V-2s launched from trains and submarines. Other V-2s, to improve their accuracy and survivability, were to be housed in hardened bunkers. Additionally, further developments of the V-2, such as the A-9 and A-10, were begun in 1941 and 1942

Left above: Wernher von Braun photographed at Peenemünde in 1944 with (left, back view) Gen Walter Dornberger and (with binoculars) Gen Schneider.
Above: The shaky liftoff of a captured 'V-2' which was fired from the stern of the carrier USS *Midway* on 6 September 1947.
Left: An RAF officer examines a captured Henschel Hs 293, a radio-guided rocket missile used with devastating effect against Allied ships from August 1943.
Below: The first US generation after 'V-2' was the Army's Redstone, for which chief contractor was Chrysler. Here one of these 69-ft missiles is pulled upright.
Right: JB-2 was an American copy of the 'V-1' cruise missile; this one was fired at Holloman Field, Albuquerque NM.

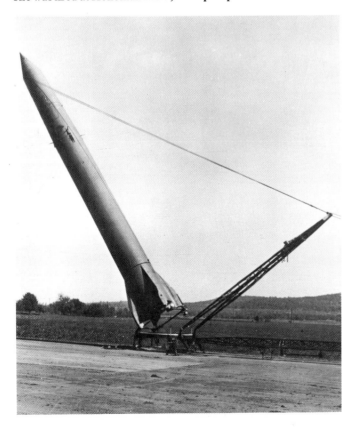

as a way to launch intercontinental strikes against North America by the late 1940s.

The last major rocket development of the Peenemünde effort for modern missiles was the Waterfall surface-to-air missile. Begun in 1942 and tested in 1944, the Waterfall SAM was to be employed with other missiles in a zone defense plan against US and British bombers in western Germany. The end of the war came first. Waterfall was unique in that its liquid propellant would always have been available for launch, unlike other missiles of the day whose fuels could not be stored indefinitely. This technology would later appear in US and Soviet missiles as a way to make them more reliable and easier to handle.

At the end of the war German scientists who had worked at Peenemünde were captured by the US and the Soviet Union. These groups would form the nucleus of missile development in both countries. Their work, of course, began with improving what had worked before – the V-2.

Some of the Sciences of World War III
Multiple Independently Targetable Reentry Vehicles

The deployment of MIRVs fundamentally enhanced ballistic-missile effectiveness by enabling a single missile to attack several different targets, or to attack a single target with more than one warhead. In either case the effectiveness of a single Intercontinental Ballistic Missile (ICBM) was significantly improved by the MIRV warhead. This has significant implications for strategic stability: for the first time, one force of ICBMs has at least the theoretical capability to undertake a disarming first strike effectively upon another ICBM force of equal or even somewhat larger size. MIRVs also increase the number of warheads that the attacker can use to saturate an ABM defense to increase the likelihood that such warheads would destroy their targets. The United States began deploying MIRVs on its land-based ICBMs in 1970; the USSR began in 1974.

Accuracy is another critical factor in determining ICBM effectiveness. It is estimated that the most accurate of the current American ICBMs, the improved version of the Minuteman III, can deliver its warheads within 900 feet of its targets. By the late 1980s, when the next-generation US ICBM could become available, it is expected that its accuracy could be improved to something less than 600 feet. This would result from a sophisticated new guidance system utilizing the advanced inertial reference sphere (AIRS). AIRS would minimize ICBM guidance errors by providing the missile with an inertial measurement unit in a 'floated ball' platform that avoids the errors resulting from the earlier gimbal-based guidance systems. This new guidance system, coupled with even a relatively modest warhead in the 150-300 kiloton range, would substantially improve US capability to destroy even extremely hard targets.

This trend toward increasingly accurate inertial guidance systems has certain significant effects. One is that as ICBM accuracy reaches a certain point – around 600 feet – the amount of throw-weight or megatonnage required to destroy most hard targets begins to become of secondary importance, because the missile's accuracy has reached a threshold where any reasonably sized warhead can destroy even a very hard target. Advanced inertial guidance systems that could be available in the late 1980s are likely to increase ICBM accuracy enough to cross that threshold.

Although the deployment of MIRVs and the improvements in ICBM accuracy have contributed substantially to the vulnerability of silo-based missiles, other weapon characteristics will become increasingly critical in determining ICBM effectiveness once a nation has acquired a certain number of highly accurate MIRV warheads. These characteristics are ICBM reliability and the ability to detonate warheads with precise timing during a large-scale counterforce attack.

Precise timing of the arrival and detonation of nuclear weapons is essential for avoiding the problem of 'fratricide,' whereby the nuclear weapon effects resulting from the initial warhead's detonation could degrade the effectiveness of the following warheads or even disable them. Following the detonation of an initial warhead over its target, there would be an opportunity of a very few seconds for a second warhead to be effectively detonated over the target. The use of a second warhead requires precise timing in arranging the arrival of the hundreds of warheads over their targets, for if the warheads in the second wave arrive too closely behind those in the first, they could be destroyed by the detonations of the first warheads, while if they arrive too late, they could be degraded by dust clouds and other indirect weapon effects created by the initial detonations. The fratricide problem is therefore believed to limit the attacker to no more than two warheads per target during a single strike, as well as requiring the precise timing of the arrival of the two waves of warheads. Although it has been pointed out that there are great difficulties inherent in attempting a nuclear strike with more than one warhead per target, recent Defense Department studies have concluded that 'with careful planning it is possible to launch a number of warheads against an individual silo without their having a mutually deleterious effect.'

Sea-Launched Ballistic Missiles

The SLBM's inability to destroy hardened targets with high reliability due to its limited accuracy had led many to see the primary role of the nuclear missile submarine (SSBN) force as a strategic reserve force targeted upon the adversary's urban-industrial centers. The mobility of the SSBN, hence the uncertainty concerning its precise launch location, has traditionally confined its role to targeting the more vulnerable soft targets, while more accurate ICBMs have significant capability against both soft and hardened targets.

Several trends in accuracy improvements, however, may give SLBMs a significant capability to destroy hardened targets by the late 1980s or early 1990s. The Trident II SLBM is expected to have significantly more throw-weight, permitting it to carry warheads with larger yields. The small yield of the current US SLBM warheads (about 40 kilotons per warhead) places great demands upon the accuracy required to destroy hardened targets. A larger yield for future US SLBM warheads would somewhat improve their capability to destroy hardened targets. Further improvement in SLBM accuracy will contribute most to hard-target lethality.

The United States has only recently begun to study seriously methods for improving the accuracy of its SLBMs. The improved accuracy program seeks a better understanding of the specific error sources that influence SLBM accuracy. Its results are expected to become the basis for upgrading Trident I accuracy and for decisions on the Trident II guidance system. Unlike land-based ICBMs, which can achieve greater accuracy with only inertial guidance, it appears that SLBMs will

Launch of one of the first American ICBMs, an Atlas F with General Electric Mk 3 reentry vehicle. The 75-ft airframe was made of thin stainless steel, stabilized by internal pressure like a balloon.

14

need more advanced systems, such as mid-course and terminal guidance systems, to achieve high levels of accuracy. Concepts now under study include very accurate in-flight updates of the SLBM's position utilizing satellite navigation systems and several types of terminal guidance sensors.

Although some say highly accurate SLBMs are likely to diminish strategic stability if deployed as a supplement to an accurate land-based ICBM force, they can, as others argue, contribute to stability by permitting the phasing out of vulnerable ICBM forces. As mentioned earlier, a unique characteristic of ICBMs has been their ability to attack time-urgent hardened targets. The deployment of highly accurate SLBMs will give the SSBN force a similar capability for the first time, enabling it to be more easily substituted for the ICBM force in the future.

SSBN Survivability
A major SSBN characteristic has been its high degree of survivability. This results primarily from its ability to

take full advantage of the concealment provided by the ocean and to remain submerged during the entire length of its patrol. Both factors reduce the probability that it will be detected. It is believed that modern SSBNs are much further ahead of the threat they presently face from various anti-submarine (ASW) forces than are ICBMs and strategic bombers. It is also generally believed that the United States has maintained an advantage over the Soviet Union in terms of ASW capability, but that neither country can successfully undertake a preemptive strike against the other's SSBNs at sea. The possibility exists, of course, that future technological developments could change this situation.

Anti-Ballistic Missile System
A ballistic missile defence (BMD) system seeks to destroy ballistic missile warheads before they reach their targets. A ballistic missile goes through several phases en route to its target, and interception at each flight phase poses different problems for the defender. The

Left: The first sub-surface launch of an early Tomahawk cruise missile, fired inside a protective capsule off San Clemente in February 1976.

Below: Following Redstone the US Army was forbidden to develop really long-range missiles and its largest weapon ever since has been the portable Pershing system. Here a Pershing 1a is being erected.

missile proceeds successively through the boost phase, in which it is propelled beyond the atmosphere; the midcourse phase, in which it can deploy penetration aids and multiple warheads to confuse or overwhelm the defense; and the terminal phase, as it reenters the earth's atmosphere. To date, US BMD systems have concentrated on intercepting the attacking warheads in that portion of the terminal phase of their flight where identification of the actual warhead is made possible because atmospheric drag filters out the lighter and slower penetration aids from the heavier and faster warheads.

Midcourse and boost-phase BMD systems have never been deployed and are apparently not currently deployable, though considerable research is underway for the former. Earlier interception is of military interest because it enables interceptors to protect much larger land areas. Midcourse interception of a ballistic missile, if achieved before the missile deploys its multiple warheads and penetration aids, enables one interceptor to perform a task that would otherwise require many single-warhead interceptors. Boost-phase interception, which would require technologies not deployable for many years, would combine the advantages of midcourse interception with the ease of destroying a comparatively slow target containing large amounts of volatile fuel.

The principal current elements of a BMD system are radars, interceptor missiles, computers and command and control systems. The BMD radars detect and track the attacking missiles, as well as guiding the defensive missiles to their intercept point. Detection and initial tracking is usually performed by special long-range early-warning radars, while the final intercept tracking and defensive missile guidance is often performed by separate shorter-range battle management radars.

Although both radars rely on automatic phased-array scanning, in the United States the long-range radar has come to be known as the PAR (perimeter acquisition radar), while the battle-management radar was designated the MSR (missile site radar).

BMD interceptor missiles destroy the attacking warheads. In the past the United States developed two such interceptors. The longer-range missile (Zeus, succeeded by Spartan) was tasked with intercepting enemy warheads beyond the earth's atmosphere (this exoatmospheric intercept can occur at about 150,000 to 400,000 feet). Later the United States developed a shorter-range hypersonic-speed interceptor, Sprint, that would attempt to destroy the remaining enemy warheads within the atmosphere just before they reached their targets (an endoatmospheric intercept, at altitudes from 30,000 to 150,000 feet). Special nuclear weapon effects, such as X-rays for the exoatmospheric interceptor and neutrons for the endoatmospheric interceptor, would be used to disable the attacking warheads.

The computer hardware and software supporting the radars and interceptors are critical elements of the BMD system. They must execute a massive number of instructions at almost real-time speeds and with almost perfect reliability. The computer systems are particularly crucial for enabling the BMD to discriminate warheads from the large amount of debris and decoys, and to plot intercept flight paths for the defensive missiles. They are the critical links between the hardware elements – the radars and the interceptors. It is advanced computer technology that allows the potential of the radar sensors to be effectively utilized to identify the enemy warheads, track them and guide the defensive missile to its intercept point.

BMD requirements change over time. A successful BMD system must not only counter the current offensive threat, but also respond to future improvements in the adversary's offensive forces. The changing and expanding strategic offensive forces of both sides have created increasingly complex requirements for BMD systems. Over the past decade, as the number of ICBMs and SLBMs deployed by each side increased, and missiles armed with multiple reentry vehicles (MRVs) or multiple independently targetable reentry vehicles (MIRVs) became available, the possibility of saturating a BMD system, thereby allowing a few warheads to 'leak' through the defenses and destroy their targets, became a serious problem.

Offensive forces can also attempt to confuse and overwhelm the defense by using various penetration aids to hide the actual warheads during a portion of their flight. These could range from such simple devices as the debris from the missile's booster and chaff (long metal strips that reflect radar waves) to more sophisticated decoys that resemble the warheads for a longer period of flight. Electronic jamming and nuclear blackout resulting from high-altitude detonations are other techniques the offense can use to degrade the defense's radar capability. Finally, a much more advanced countermeasure to a ballistic-missile defense system would be the deployment of a maneuvering reentry vehicle (MARV), which would change its course during the terminal phase of its flight in order to prevent the defense from accurately calculating its trajectory.

BMD systems are of particular interest for performing two types of missions, city defense and ICBM defense. Many view ICBM defense as nearly synonymous with point defense. Others consider city defense to be a form of point defense, and area defense to be protection of vast geographic areas, tens of thousands of square miles. City defense has stringent technical requirements. The defense must be perfect, as one large nuclear weapon would wreak havoc on any city. The defense must have enough interceptor missiles to destroy large numbers of reentry vehicles at each city. The defense must protect a large area, as a large nuclear explosion even miles away could damage the city. Such a defense is extremely difficult, especially one using ABM interceptor missiles to defend against a sophisticated Soviet attack.

Defense against a light non-Soviet attack is easier, but has not been a major issue for several years. In contrast, defense of ICBM silos is much simpler and much cheaper. Since these silos are hardened against nuclear attack, only a near-miss or a direct hit can destroy the missiles they contain. The defense need not be perfect. Since some losses are expected and permissi-

A view of Nike Hercules missiles on their launchers.

ble, ICBM defense can rely on the atmosphere to help separate real warheads from decoys.

The Future
As seen below, there are a variety of ways to exploit technology in missile construction and propulsion: how a missile is launched and controlled once it is in the air as well as how it is guided to its target. Furthermore, there are several different modes for delivering nuclear warheads.

Missile advances are being made regularly and future missile capabilities, because of various technologies, are likely to be even more exotic. 'Invisible missiles,' terminal guidance systems, improved propulsion and the introduction of non-nuclear warheads in strategic weapons will all have impact on the future weapons of World War III. Nuclear warhead development for mis-

siles is also likely to evolve along various paths. The first nuclear warheads were developed in the late 1940s and 1950s. They were large in both size and yield of the explosive. By the late 1950s and 1960s nuclear warheads were reduced in size without loss of yield. In the future the size of the warhead will not be an issue. A future warhead will have as options, through progress in computers and physics, a selection of those effects of a nuclear explosion which it will produce. Consequently, depending on the target, either the heat, radiation or blast effects of a nuclear explosion (or combinations) will be exploited, increasing the military utility of the missile.

The technical evolution and sophistication now found in modern strategic weapons has ensured that missiles will continue to dominate the offensive aspects of any future war.

A Jupiter missile blasts skyward in a spectacular test
launch from Cape Canaveral.

2. US STRATEGIC MISSILE FORCES

Out of World War II came jet planes and modern missiles and rockets. US experience in that war, though, favored long-range strategic bombing from manned aircraft as a means to strike its new adversary – the Soviet Union.

Any initial interest in developing long-range ballistic missiles would be confined to research and development studies. Air-breathing missiles such as the V-1 were at the time believed to be less risky to develop and operate. As a result efforts, although relatively indifferent compared to bomber development, were expended on both bomber-launched cruise missiles and cruise missiles that could be fired from bases on land.

By the end of the Korean War a number of political and military events had changed. The Soviet Union, through its military activity in Eastern Europe and because of its nuclear testing, was the new military threat. On the technical side, improvements in warhead packaging, guidance systems and propulsion technology made long-range ballitic missile development more attractive. Consequently, there were now urgent political and military requirements to develop ballistic missiles. As a result, by 1954 a national effort, rather than an Air Force effort, was instituted to build ballistic missiles quickly.

For the next thirty years ballistic missile develop-

ment would progress through the regular introduction of regional and intercontinental range, liquid and solid propellant, land- and sea-based ballistic missiles. The regularity of the introduction of missiles that were to fight a future war was of course tempered by numerous domestic, political, and bureaucratic struggles, as well as varied interpretations of the Soviet and Chinese military threat. Which missiles, and how those missiles satisfied those political and military requirements are described below.

The Post-War Planning Environment

Less than one year after the close of World War II the Strategic Air Command was established as one of the three major combat commands of the US Army Air Force. Its mission was to conduct long-range offensive operations into any part of the world.

As a result of the importance of the strategic bomber in World War II, Air Force attention and development work was geared toward supporting and enhancing the manned bomber force rather than toward building ballistic missiles. The major missions of the strategic bomber in World War II were to destroy Germany's war-making capacity and submarine construction yards. In the post-war world the major target of the United States was the Soviet Union and its ground and

Strategic Intercontinental Forces							
	1950	1955	1960	1965	1970	1975	1980
Bombers							
B-29	286	–	–	–	–	–	–
B-50	196	–	–	–	–	–	–
B-36	38	205	–	–	–	–	–
B-47	–	1086	1178	114	–	–	–
B-52	–	18	538	600	459	420	362
B-58	–	–	19	93	–	–	–
FB-111	–	–	–	–	42	69	66
ASM's[2]	–	–	54	542	345	1759	1020
Subtotal (aircraft)	520	1309	1835	807	501	489	428
Subtotal (bombs and missiles)	68	2310	4288	2772	1972	2790	2412
Tankers							
KB-29	126	82	–	–	–		–
KC-97	–	679	689	–	–	–	–
KC-135	–	–	405	615	615	599	515
Subtotal	126	761	1094	615	615	599	515
Land-Based Missiles							
Snark	–	–	30	–	–	–	–
Atlas	–	–	12	–	–	–	–
Titan I[3]	–	–	–	–	–	–	–
Titan	–	–	–	54	54	54	53
Minuteman I	–	–	–	800	490	–	–
Minuteman II	–	–	–	–	500	450	450
Minuteman III	–	–	–	–	10	550	550
Subtotal (launchers)	–	–	42	854	1054	1054	1053
Subtotal (warheads)	–	–	42	854	1069	1879	1878
Sea-Based Missiles[4]							
Polaris A-1	–	–	32	80	–	–	–
Polaris A-2	–	–	–	208	128	48	–
Polaris A-3	–	–	–	208	512	208	160
Poseidon C-3	–	–	–	–	16	400	480
Trident C-4	–	–	–	–	–	–	16
Subtotal (launchers)	–	–	32	496	656	656	656
Subtotal (warheads)	–	32	32	496	768	3456	4128
Total Launchers	646	2070	2903	2772	2826	2798	2612
Total Warheads	68	2310	4362	4122	3809	8123	8418

[1] Not including those in storage. [2] Air-to-surface missiles such as Hound Dog and SRAM.
[3] In service between 1961 and 1964. [4] All deployed aboard 16-tube SSBNs.

While development Snark intercontinental cruise missiles were red, those used operationally by USAF Strategic Air Command were gray. Here an SM-62 Snark is towed to the launcher.

air forces facing western Europe. Consequently US strategic requirements focused on striking the Soviet Union's industrial heartland to disable the Soviet's war-making effort.

This mission, for which some 50 nuclear weapons were available in the late 1940s to strike about 70 urban industrial targets, began to change as a result of a number of political and military events. These included the Soviet blockade of Berlin, the Soviet testing of atomic weapons in 1949, Mao Tse Tung's victory in the Chinese civil war (and his alignment at that time with the USSR) as well as the start of the Korean War, which some believed to be the start of a Third World War that would eventually include Europe.

As a result, force targeting requirements were to include the destruction of the Soviet Union's war-making potential as well as direct nuclear attacks on its ground and air forces before they could strike at Europe or the United States. In this manner US strategic forces provided extended deterrence to its allies.

To support these evolving targeting requirements, the US Air Force, as well as the US Navy, developed a number of weapon systems that could carry nuclear weapons over strategic distances. Some were to be launched from bombers, submarines or aircraft carriers while others were to be fired from launch ramps on land bases.

The Bomber Legacy

After World War II 25 German V-2 rockets were assembled from spare parts brought to the United States from Peenemünde. Despite successful testing the future of strategic air power was still believed to be in the nuclear-armed long-range jet bomber. To complement the bomber, aerodynamic cruise missiles fired from the bomber were developed as well as long-range unmanned aerodynamic missiles.

At the time the jet bomber and the cruise missile seemed best to satisfy the weapons requirements for World War III. There were considerable pressures and political battles to maintain the bomber and the nuclear bomb as the preeminent weapons system. The Air Force was already struggling with the Navy over future roles and missions. In addition the technology for jet bombers and aerodynamic missiles was proven, while that for long-range ballistic missiles was not. For example, in the 1940s and early 1950s atomic weapons were large and heavy. Attaching them to ballistic missiles that were still being tested only at short ranges was, at best, a high-risk option.

Between 1946 and the end of the Korean War some ten aerodynamic missiles began serious development. In addition one ballistic missile, the Redstone, also began development directed by Wernher von Braun and his V-2 missile design team. Of these eleven weapons, nine had been tested by 1956 and had been deployed between 1954 and 1958.

The Post-War Missiles

The Snark intercontinental cruise missile was designed by Northrop Aviation. Development began in 1946 (as the MX-775), the first test flight occurred in 1953 and deployment began in 1957. The missile was powered in subsonic flight by a turbojet engine after being boosted into flight by two solid-propellant rocket motors. The Snark used stellar inertial guidance and could attack the Soviet Union from a variety of azimuths on its 10-12 hour mission. Thirty Snarks were based at Presque Isle Air Force Base, Maine, each armed with a 5-megaton (MT) warhead.

The Rascal supersonic air-to-surface missile was designed by Bell Aircraft. Development began in 1946, the first test flight occurred in 1953 and the missile became operational in 1957. The rocket-powered command-guided weapon was an attempt to reduce bomber vulnerability to Soviet air defense. The missile was tested by the B-36 and B-47 but was operational only on the B-47.

The Navaho was a supersonic intercontinental aerodynamic missile designed by North American Aviation. Development began in 1947 in place of funds for ICBM research, and after a long development period the missile was tested and canceled in 1957 and 1958. The development of the Navaho provided much of the technological foundation for future ICBM work on Jupiter, Thor, Atlas and Titan I. This included engine, propellant and guidance components. Navaho was powered in vertical takeoff from its launch pad by three rocket engines producing over 200 tons of thrust which, combined with twin turbojets for cruising, powered this bomber-sized missile at better than Mach 3.0. The Navaho was guided on its three-hour intercontinental mission by an advanced inertial guidance system and may have been scheduled to carry a 24 MT warhead.

The Rigel supersonic cruise missile built by Grumman was to be an offshore bombardment missile launched by a submerged submarine. Rigel development began in 1947; it was tested in 1951 and canceled in 1952 in favor of the Regulus I.

The Regulus I subsonic cruise missile built by Chance Vought began development in 1947. It was first tested in 1951 and entered service with the Navy. It was powered by a turbojet engine and guided by radio command, designed to carry a high-explosive warhead or 400 kiloton (KT) warhead over 375 nautical miles (NM). The Regulus I was an attempt by the Navy to diversify and supplement its carrier-based aviation with missile power and as a way to minimize aircraft losses during combat operations. Some 157 launchers were in service at various times on five submarines, four cruisers and ten aircraft carriers.

The Matador was a subsonic aerodynamic missile built by the Martin Corporation. Development began in 1947, it was tested in 1953 and entered service with the Air Force in 1954. It was powered by a turbojet engine after being put into flight by a rocket booster. Command guidance steered the missile over its 650-NM course. Matador supplemented the USAF's Tactical Air Command's tactical nuclear interdiction capability with a 200 KT warhead. Consequently, 72 Matador launchers were based in Germany starting in 1954 while 24 launchers were based in Korea. An additional 48 launchers were deployed to Okinawa. All were co-located with other Air Force units and were phased out of service in 1962.

The Redstone was at the foundation of the US

Above: The big USS *Grayback* (redesignated LPSS-574 from SSG-574) was the first to fire the Regulus II missile in September 1958. It could fly 1,000 miles at Mach 2.
Left: USS *Tunny* (SSG-282) ready to fire a Regulus I cruise missile of the US Navy on 26 August 1954. Missile range was 400 miles.

Above: Artist's impression of the USAF Fairchild XSM-73 Goose, a cruise missile launched from hardened shelters.

Below: Artist's impression of launch of a USAF Martin TM-76 Mace cruise missile from its hardened shelter. Another variant was fired from a cross-country mobile launcher.

Army's ballistic missile effort following World War II and was an evolutionary V-2 design. The missile was developed by the Redstone Arsenal under the guidance of Wernher von Braun's design team in 1950. Redstone was first flown in 1953 and became operational in 1958. It was a liquid-fueled missile but was another beneficiary of the propulsion research done for the Navaho missile. Guidance was inertial. Redstone had a 250-NM range with a 300 KT warhead. Sixteen mobile launchers were deployed to Europe as an extension of the VII Army Corps artillery. The missile was phased out of service as easier-to-handle solid-propellant missiles became available.

The Goose was an aerodynamic cruise missile developed by Fairchild, intended to confuse enemy radar during bomber raids. Development began in 1951 and preliminary flight testing started in 1957. The missile was canceled in favor of Snark before long-range flight testing started.

The Triton was an aerodynamic cruise missile built for the Navy by the Applied Physics Laboratory at Johns Hopkins University. It began development in 1951 but was canceled in 1955. It was to be powered by an integral rocket ramjet over a distance of 375 NM. It was also to be directed by an advanced guidance system made up of radio command, inertial and a map-matching targeting system. Its warhead was to be in the 1-10 KT range. The system had other features such as folding fins for submarine storage.

The Regulus II was a supersonic aerodynamic cruise missile built by Chance Vought that began development in 1953. It was flight tested in 1956 and became operational in 1958. Supersonic speed was achieved through advanced aerodynamics and turbojet power. Guidance through its 1000-NM flight was fully inertial, and it carried a four-MT thermonuclear warhead. Regulus II was an advanced land-attack strategic missile whose range speed and payload equaled that of manned aircraft then operational. Its range, unlike earlier naval missile systems, gave the Navy some flexibility and increased survivability in wartime operations around the Soviet Union and China. Before Regulus II was retired from service in 1959, because of Navy ballistic missile development, it was in operation with five submarines and one cruiser.

The Mace was a direct product improvement of the Matador. Design work by Martin began in 1953 and it was test fired in 1955. Deployments began in 1959 as a replacement for Matador. While having similar rocket boost and turbojet transonic cruise technology, the larger size of Mace increased the weapon range to 1100 NM. Guidance was a combination of radio command and a radar-matching terrain map system. Mace is also believed to have had a four-MT thermonuclear warhead, giving the Air Force an extremely powerful theater attack weapon. Some 144 Mace launchers were deployed with the 17th Air Force in Germany beginning around 1959. Forty-eight other Mace launchers were based in Okinawa with the Pacific Air Forces. All were located in above-ground shelters, adding some protection against attacks. Mace units were withdrawn from service throughout 1968 and 1969.

US Missiles in the Era of Massive Retaliation

By 1954 the Korean War was over and the Soviet Union had tested its first thermonuclear weapon. The Eisenhower administration had also been in power one year and was addressing its dissatisfaction with the doctrines and strategy of the post-war years, particularly the limited war in Korea and the small size of the defense budget prior to that conflict. In January of 1954 the Secretary of State, John Foster Dulles, announced a defense strategy called Massive Retaliation. Together with a foreign policy of containment that also included Collective Security arrangements with allies, Massive Retaliation as a declaratory policy threatened the Soviet Union and China with overwhelming military force at a time and place of US choosing if Soviet or Chinese forces attacked Western interests. Massive Retaliation was also contingent on nuclear superiority.

One by-product of the change in general strategy was a greater interest by US planners in targeting Soviet and Chinese military forces, rather than concentrating on the urban industrial centers of the Soviet Union and China. Success in this type of 'spasm war' would be achieved by striking enemy conventional and nuclear forces. Military forces were, of course, not yet available in 1954 for these missions. Further, reconnaissance of the Soviet Union and China was probably inadequate for the types of targeting the services would be required to do in support of Massive Retaliation.

Consequently, the first nuclear forces deployed to support a Massive Retaliation doctrine were those forces that initially began research and development in the post-war period and were entering service in 1954. These included such theater-strike systems as Regulus I and Regulus II, Matador and Mace, Redstone, the Rascal bomber-launched air-to-surface missile and the intercontinental-range Snark cruise missile. These missiles, combined with a modern bomber force, offered diverse capabilities for US planners against enemy theater forces as well as his urban industrial base. Nevertheless, missiles had not yet been fully accepted into strategic force planning. That would come, as will be seen with the development of ballistic missiles and thermonuclear weapons.

While work on aerodynamic missiles and jet bombers progressed rapidly following World War II, ballistic development lagged. Of course there was greater Air Force and Navy interest in such established programs as manned aircraft launched from land bases as well as aircraft carriers. But, in part because of World War II's end and defense budget decreases, a number of aviation companies began to take a greater interest in research and development of missiles. Early technical requirements specified a number of different missile projects of different ranges.

Developing a ballistic missile capable of matching medium bombers (B-47) and heavy bombers (B-52) that were already on the drawing board would be a difficult task. A V-2 missile, to compete with a bomber, would have had to increase its targeting accuracy from five miles to one-half a mile. Its range also would have to be increased from 300 NMs to 1500 NMs if the missiles were to be based in Europe, or to 6000 NMs if the missiles were based in the United States.

The initial intercontinental ballistic missile research project was the Consolidated Vultee project MX-774, a V-2 outgrowth, that began funding in 1946 (Consolidated Vultee would later become Convair). The MX-774 contract laid the foundation for future ballistic missile development and would lead directly to the Atlas ICBM. In 1947, however, a small Air Force budget canceled the project. Despite the cancellation and test failures in 1947 and 1948, various advances in missile technology were made. MX-774 was funded by Convair through 1950, and in 1951 the Air Force issued a second contract to Convair for ICBM work. This was to be MX-1593, the blueprint for the Atlas ICBM.

The MX-774 was a major step forward, incorporating a number of features that distinguished it from the German V-2. If development of the MX-774 had been fully funded, expanded and supported from the start, SAC could have had an operational ICBM as early as 1955. There were three innovative features of the MX-774. They included gimbaled (swiveling) engines to deal with missile flight direction in place of the power-robbing vane deflection system of the V-2; propulsion concepts that eliminated internal fuel tanks and used the walls of the missile instead (as well as using nitrogen gas to support the missile structure rather than metal supports as on the V-2) and, finally, the introduction of a separate warhead stage rather than having the entire missile impact.

The second effort to develop an ICBM (MX-1593) might also have failed to attract support had it not been for the thermonuclear hydrogen bomb breakthrough reached in 1952. With this achievement, ten times more nuclear power could be contained in a single bomb. Early missiles like Atlas would now be able to attack a target with three megatons of nuclear destruction rather than 'only' 300 kilotons. The difference was important. Now a ballistic missile would not have to have large and powerful propulsion systems to lift a large atomic payload, nor would it need the precise accuracy of a manned bomber if it were a medium-size missile carrying a small or medium atomic weapon. Consequently, the hydrogen bomb breakthrough brought ballistic missile development to the point where it was both technically achievable and militarily desirable.

By 1954 the Eisenhower administration was convinced by numerous groups and committees to begin a crash ICBM program. The end product was the Atlas ICBM, built under the enlightened direction of Air Force General Bernard A. Schriever.

Other ballistic missile systems also came into the concept phase. These included another Air Force ballistic missile of intermediate range (Thor), as well as a joint Army and Navy intermediate-range ballistic missile (Jupiter and later Polaris).

Because of multi-service interest in missile development, for both budgetary and tactical reasons, various struggles emerged between the services as to which would control the new missiles or whose missiles would be procured for service. Because the Air Force, with its strategic bombers, already had an intercontinental role and was far along in its development of Atlas, the

Left: A development SM-78 Jupiter of the US Army at liftoff from Cape Canaveral on 17 July 1958.
Above: Atlas ICBM No 134F flew the slim Chrysler ABRES re-entry vehicle on 1 March 1963.
Below: Atlas ICBMs and space launch vehicles undergoing modification at Edwards Air Force Base.

mission to strike the Soviet Union with missiles in a future World War would remain an Air Force prerogative. The Army would retain missiles to support its ground forces in Europe and Asia. Accordingly, the Jupiter IRBM was transferred from the Army to the Air Force and was deployed overseas with Thor.

The testing of the first Soviet ICBM accelerated missile development in the United States. But by 1957 a number of ballistic missile projects were already moving forward. These included Atlas, Jupiter, Thor, the Titan I, Polaris, Minuteman I and Pershing I. In addition, supersonic aerodynamic bomber missiles such as the Hound Dog were also underway. These systems would be deployed between 1958 and 1962, reflecting the political and military urgency that existed at the time. After 1957, and through the end of the Eisenhower administration, two more weapon developments began in the United States. These were the Titan II ICBM and the Skybolt bomber-launched ballistic missile. They would also be ready for deployment after 1962.

Weapon System-107A (WS-107A) became in operational form the Atlas ICBM. Built by Convair as a crash program beginning in 1954, the Atlas was first tested to full range in 1958 and became operational in 1960. Atlas was a multi-stage liquid propellant missile, first with radio-inertial guidance (Atlas D) that prohibited salvo launching and later (Atlas E and F) with standard inertial guidance equipment. Missile accuracy was about 2.0-NM circular error probable (CEP), carrying a three-megaton warhead over great distances even for ICBMs. As a result, Atlas would have been used against airbases and industrial complexes in the Soviet Union and China. By program's end 126 Atlas missiles of various types (Atlas D, E and F) were deployed with 13 SAC missile squadrons at 11 airbases in the United States. Some of these missile bases were located in the southern United States, taking advantage of Atlas' great range.

Atlas basing was a mix of above-ground launch pad complexes, coffin fixtures and silos. Some 36 Atlas D missiles were on above-ground pads and launch preparations took more than one-half hour. The Atlas E, of which there were about 18, lay horizontal in a steel coffin and had to be raised into firing position. This shelter would have provided some minimal blast protection. The Atlas F, of which there were about 72, was stored in first-generation underground silos, giving added protection against nearby nuclear explosions.

The Jupiter was designed under the aegis of Wernher von Braun as an intermediate-range ballistic missile follow-on to the Redstone short-range ballistic missile. Missile design work began in 1954 with the first flight test taking place in 1957. The missile became operational in 1959 and remained in service until 1965. Jupiter carried liquid propellant, had inertial guidance (a 1.0-NM CEP) and carried a one MT warhead. Forty-five Jupiters were deployed overseas to be operated by the Italian Air Force (30) and the Turkish Air Force (15).

Jupiter had a checkered military career. Originally the missile was to be an Army-operated strategic weapon, but by 1955 the Eisenhower administration

Top: A sequence showing a rare silo launch of a Titan 1 ICBM; this missile was normally hoisted to the surface first.
Second above: Removing a Titan 1 ICBM from its deactivated silo in 1965.
Left: Douglas broke all records in their quick development of the SM-75 Thor IRBM; this was the first to fly.
Above: The heavy copper heat-sink nosecone for a Thor IRBM.

Summary: United States

Missile	Designer	Year Design Began	First Flight Test	Propulsion System	Guidance	Warhead	Range	Year Operations Began	Weight at Liftoff	Basing Mode	Number Deployed
SM-62 Snark Cruise Missile	Northrop	1946	1953	Turbojet	Celestial Inertial	One 5 MT	6250 NM at Mach 0.90	1957 to 1961	30 tons	Ramp	30
SM-65 CGM-16; HGM-16 Atlas ICBM (formerly WS-107A1)	Convair	1954	1958	Liquid fuel	Radio Inertial	One 3 MT	9000 NM	1960 to 1967	130 tons	Above-ground Pad and Shelter	120
SM-68; HGM-25A Titan I ICBM (formerly WS-107A2)	Martin	1955	1959	Liquid fuel	Radio Inertial 0.75 NM CEP	One 4 MT	8000 NM	1962 to 1966	110 tons	Silo Above-ground launch	54
HSM-80A; LGH 30AB Minuteman I ICBM (formerly WS-133A)	Boeing	1957	1959	Solid fuel	Inertial 1.0 NM CEP	One 1.3 MT	7000 NM	1962	32 tons	Silo	800
SM-68B; LGH-25C Titan II ICBM	Martin	1958	1961	Storable liquid	Inertial 0.70 NM CEP	One 7.4 MT	9000 NM	1963	165 tons	Silo	54
LGM-30F Minuteman II	Boeing	1962	1964	Solid fuel	Inertial 0.25 NM CEP	One 1.1 MT	7000 NM	1966	35 tons	Silo	500
LGM-30 G Minuteman III ICBM	Boeing	1964	1968	Solid fuel	Inertial 0.15 NM CEP	MIRV 3×170 KT and 3×335 KT	7000 NM	1970	38 tons	Silo	12 / 300
Peacekeeper (MX)	Boeing	1974	1983	Solid fuel	Inertial AIRS 0.05 NM CEP	MIRV 10× 335 KT	7000 NM	1986	95 tons	Dense-Pack Silos?	100
SM-78; PGM-19A Jupiter IRBM	Redstone Arsenal	1954	1957	Liquid fuel	Inertial 1.0 NM CEP	One 1 MT	2000 NM	1958 to 1965	55 tons	Above-ground Pad	45
SM-75; PGM-17A Thor IRBM (formerly WS-315A)	Douglas	1955	1959	Liquid fuel	Inertial 1.0 NM CEP	One 1 MT	2000 NM	1959 to 1965	53 tons	Above-ground Pad	60

had decided that a sea-going Jupiter should be developed by the Navy as well. This would soon lead to Polaris. As the missile was to begin flight testing, the Eisenhower administration limited all Army ballistic missiles to under 200 miles. As a result, the Air Force took over the project. The main feature of the Army Jupiter was to be its mobility, similar, in many respects, to the V-2. But once under Air Force control, all mobility aspects of the missile were removed; like other Air Force missiles of the time, it was based in above-ground launch pads.

In operational colors the WS-107A-2 became the Titan I ICBM, designed by the Martin Corporation beginning in 1955. It was first flight tested in 1959 and was operational from 1962 to 1966. Like Atlas and Jupiter, Titan I contained liquid propellant (liquid oxygen and kerosene) that was loaded prior to firing. Titan I had radio-command inertial guidance with a 0.75-NM CEP. Titan had a four-megaton warhead and 54 were deployed in the United States at five bases. Titan I was housed in silos, although it still had to be raised to the surface to fire; reaction time was on the order of 20 minutes. Titan I was developed as a fall-back if Atlas failed. It was more sophisticated because, unlike Atlas, its second-stage rocket motor turned on only when the first-stage motor was shut off and discarded.

The Thor was built by Douglas Aircraft as an intermediate-range ballistic missile starting in 1955. It was flight tested in 1959 and became operational the same year. The timetable followed by the Thor stands as a record for major weapon system development. The Thor carried liquid propellant and was fired from above-ground launch pads. The Thor had inertial guidance providing accuracy of about 1.0-NM CEP with a one-megaton warhead. Sixty Thors were deployed in Britain with Royal Air Forces' Bomber Command. Thor was developed as an Air Force priority project similar to Atlas, although the Army and Navy already had their joint Jupiter IRBM project underway.

The Hound Dog supersonic-bomber-delivered aerodynamic missile entered the development phase in 1955. Built by North American, it was first flown in 1959 and entered service with the SAC B-52 bomber force in 1961. Nearly 600 Hound Dogs were delivered over the next two years. The Hound Dog was powered to its target at Mach 2.2 by a turbojet engine and was guided to a 3.0-NM CEP by an inertial navigation system. Hound Dog carried a four-megaton warhead. The success of the Hound Dog program led to the cancellation of the B-47 Rascal ASM program. B-52s carried two Hound Dogs. They were used primarily to strike such enemy air defense targets as air bases and SAM sites and thus assist the bomber in its penetration to its primary target.

The Polaris System

One of the more remarkable missile developments of the mid-1950s was the Polaris ballistic missile system and the fleet ballistic missile submarine that would launch it. Interestingly, the notion of a submarine-launched ballistic missile goes back again to the German V-2 research group at Peenemünde. Members of this group experimented with small rockets launched from submarines in the Baltic but, because of the pressing needs for torpedo attack submarines in the Atlantic, the German High Command terminated the project.

Various projects to develop naval missiles in the post-war period were introduced (Regulus I and others) but all had their limitations. Beginning in 1955, however, the Army and Navy entered a joint development project called Jupiter. This IRBM would be developed by the Army while the Navy worked on the launcher. In response to this, the Navy created a Special Projects Office, which, in turn, also began to work on the first solid-propellant ballistic missile for use especially in submarines, because liquid-propellant missiles like Jupiter could be a safety hazard aboard a submarine. By 1957 the Navy had withdrawn from the Jupiter project to begin working on a solid-propellant IRBM capable of being launched while submerged from the Navy's new family of nuclear-fueled submarines. The Fleet Ballistic Missile concept had two major operational advantages not shared by other systems of the time. First, the sea-launched ballistic missile, with its solid propellant, would be ready to launch on a moment's notice rather than having to be fueled prior to launch, and second, its 'launcher,' once at sea, was mobile and virtually undetectable, making a preemptive strike by the Soviet Union impossible.

The initial target date for the FBM project, including the missile Polaris, was to be 1965. But with the Soviet SS-6 launched in 1957, as well as the Sputnik launching that same year, the Special Projects Office under Admiral Raborn moved the target date ahead to 1963 and then to 1960, when it was discovered that a slight reduction in range for the first-generation Polaris would be more than worth the development time saved. Later-model Polaris missiles would make up any range deficiencies.

The Polaris A-1 missile built by Lockheed entered the design concept phase in 1955. Polaris A-1 was flight tested in 1958 and entered service in 1960. Polaris was a two-stage ballistic missile powered by solid-propellant rocket motors and had a range of 1200 NM. It was guided by a self-contained inertial guidance system. Polaris A-1 remained in fleet service through 1965; it carried a single warhead of 800 kilotons yield and had an accuracy of 0.75-NM CEP.

Polaris A-2 began development in 1959 and was flight tested in 1960. The A-2 entered service in 1962. The Polaris A-2 was 30 inches longer than the A-1 and in addition used a more powerful solid propellant. As a result its range was 1300 NM. Innovative features included second-stage rocket motor manufacture from wound fiber glass in place of steel and rotating nozzles in the motor. The A-2 had a warhead yield of 800 kilotons and accuracy of 0.50-NM CEP.

FBM Submarines

There are three classes of US Navy Fleet Ballistic Missile submarines:
George Washington Class: about 389 feet long and about 5900 tons.
Ethan Allen Class: about 410 feet long and about 6900 tons.
Lafayette Class: about 425 feet long and about 7000 tons.

George Washington Class
USS *George Washington* (SSBN-598)
USS *Patrick Henry* (SSBN-599)
USS *Theodore Roosevelt* (SSBN-600)
USS *Robert E. Lee* (SSBN-601)
USS *Abraham Lincoln* (SSBN-602)

Note: These five first carried the Polaris A-1. All were later overhauled to carry the Polaris A-3.

Ethan Allen Class
USS *Ethan Allen* (SSBN-608)
USS *Sam Houston* (SSBN-609)
USS *Thomas A. Edison* (SSBN-610)
USS *John Marshall* (SSBN-611)
USS *Thomas Jefferson* (SSN-618)

These *Ethan Allen* Class deployed carrying the Polaris A-2 missile. They were later reconfigured to fire the Polaris A-3.

Lafayette Class
USS *Lafayette* (SSBN-616)
USS *Alexander Hamilton* (SSBN-617)
USS *Andrew Jackson* (SSBN-619)
USS *John Adams* (SSBN-620)
USS *James Monroe* (SSBN-622)
USS *Nathan Hale* (SSBN-623)
USS *Woodrow Wilson* (SSBN-624)
USS *Henry Clay* (SSBN-625)
USS *Daniel Webster* (SSBN-626)
USS *James Madison* (SSBN-627)
USS *Tecumseh* (SSBN-628)
USS *Daniel Boone* (SSBN-629)
USS *John C. Calhoun* (SSBN-630)
USS *Ulysses S. Grant* (SSBN-631)
USS *Von Steuben* (SSBN-632)
USS *Casimir Pulaski* (SSBN-633)
USS *Stonewall Jackson* (SSBN-634)
USS *Sam Rayburn* (SSBN-635)
USS *Nathanael Greene* (SSBN-636)
USS *Benjamin Franklin* (SSBN-640)
USS *Simon Bolivar* (SSBN-641)
USS *Kamehameha* (SSBN-642)
USS *George Bancroft* (SSBN-643)
USS *Lewis and Clark* (SSBN-644)
USS *James K. Polk* (SSBN-645)
USS *George C. Marshall* (SSBN-654)
USS *Henry L. Stimson* (SSBN-655)
USS *George Washington Carver* (SSBN-656)
USS *Francis Scott Key* (SSBN-657)
USS *Mariano G. Vallejo* (SSBN-658)
USS *Will Rogers* (SSBN-659)

The first eight submarines of the *Lafayette* Class deployed carrying the Polaris A-2. The remaining 23 carried the A-3.
All 31 were later modified to carry the Poseidon C-3.
A further 12 were returned to fire Trident.

Polaris Missile Reliability

Launching of Polaris missiles from nuclear-powered submarines at Cape Kennedy are considered near-operational tests of the entire system, including the crew. Each submarine goes through Demonstration and Shakedown Operations (DASO) off Cape Kennedy, in which each of the SSBN's two crews is qualified to fire tactical missiles (less warhead). DASO firings are accomplished by each SSBN before beginning its initial operational patrol and after each overhaul.

DASO Results of the Three Missiles

    A-1: 21 successful out of 36 attempts = 58%
    A-2: 38 successful out of 43 attempts = 88%
    A-3: 44 successful out of 46 attempts = 96%
Between July 1964 and April 1966 there were 30 successful A-3 DASO shots in a row.

Missile	Designer	Year Design Began	First Flight Test	Propulsion System	Guidance	Warhead	Range	Year Operations Began	Weight at Liftoff	Basing Mode	Number Deployed
Rigel	US Navy	1947	1951	Ramjet	Radio Inertial	–	560 NM	Canceled	13 tons	Sub	–
Triton XSSM-N-4	US Navy (Applied Physics Lab; Johns Hopkins University)	1951	Canceled in 1955	Integral Ramjet	Inertial Radio Map Matching	–	375 NM	Canceled	10 tons	Sub	–
Regulus I SSM-N-8 RGM-6A, B	Chance Vought	1947	1951	Turbojet	Radio Command	One 400 KT	400 NM	1954 to 1964	7.5 tons	Naval	157
Regulus II SSM-N-9	Chance Vought	1953	1956	Turbojet	Inertial	One 4.0 MT	1000 NM at Mach 2.0	1958 to 1959	15 tons	Naval	6
Polaris A-1 (UGM-27A)	Lockheed	1955	1958	Solid fuel	Inertial 0.75 NM CEP		1200 NM	1960 to 1965	14 tons	SSBN	
Polaris A-2 (UGM-27B)	Lockheed	1959	1960	Solid fuel	Inertial 0.50 NM CEP		1300 NM	1962 to 1974	15 tons	SSBN	
Polaris A-3 (UGM-27C)	Lockheed	1960	1962	Solid fuel	Inertial 0.40 NM CEP	MRV	2550 NM	1964 to	18 tons	SSBN	
Poseidon C-3 (UGM-73A)	Lockheed	1961	1965	Solid fuel	Inertial 0.25 NM CEP	MIRV	2550 NM	1971	33 tons	SSBN	
Trident I C-4 (UGM-93A)	Lockheed	1971	1977	Solid fuel	Inertial plus Stellar 0.10 NM CEP	MIRV/ MARV	4000 NM	1979	37 tons	SSBN	
Tomahawk	General Dynamics	1974	1976	Turbofan	TERCOM 0.05 NM CEP	One, HE Nuclear	1500 NM	1983	1.5 tons	Naval	
Trident II D-5	Lockheed	1983	1987	Solid fuel	Inertial plus Stellar Terminal 0.05 NM CEP	MIRV/ MARV	6000 NM	1989(?)	–	SSBN	

Above: USS *Ethan Allen* (SSBN-608) was the first submarine designed from the start to carry ballistic missiles.
Left: Loading protective tubes containing Polaris A-3 missiles into USS *Stonewall Jackson* (SSBN-634) in April 1965.

The Polaris A-3 entered the development phase in 1960 and was tested in 1962; it entered service in 1964. Unlike its two predecessors, Polaris A-3 was nearly (85 percent) a new missile. In addition to possessing greater range (2550 NM), it also had the first operational Multiple Re-entry Vehicle (MRV) designed to increase the lethal blast area (three 200-kiloton warheads) or saturate ABM defenses. The accuracy of the Polaris A-3 was .40-NM CEP.

Left: Test flight of an early Boeing XSM-80 Minuteman I.
Above: Loading a Minuteman ICBM into a silo at Vandenberg
AFB for a USAF training launch.
Above right: Minuteman MIRVs streak towards Kwajalein Atoll
after launch from Vandenberg AFB, California.

The Minuteman System

Spurred by the Navy decision to attempt a solid-propellant SLBM, the Air Force began solid-propellant missile research. Up until then the Air Force appeared to be committed to large, nonstorable liquid-propellant missiles that were not only unresponsive in launch reaction time, but were exposed (because of fueling requirements) to enemy attack. Between 1956 and 1957 the Air Force studied various ways to introduce solid-propellant missiles into the force. This initially included an IRBM project as a potential Thor/Jupiter replacement and finally WS-133, the three-stage ICBM project named Minuteman. Boeing Company began formal design initiation in 1958. During the system workup to the first long-range flight test in 1961, numerous breakthroughs occurred in airframe, warhead and propulsion technology, so that when the Minuteman I became operational at Malmstrom Air Force Base in 1962 it was an extremely powerful missile, equal and superior to those many times its size.

Also unique to Minuteman was its basing plan. Rather than using above-ground launchers, as did Atlas and Thor, Air Force Minuteman planners, who were already building deep hardened-concrete silos for Titan, took advantage of the missile's small size and solid fuel and offered two adaptations, both with increased force survivability in mind. The first was a Mobile/Modular concept where different preassembled rocket stages (any combination of three) would be flown to worldwide locations to form various-range ballistic missiles. This idea was never developed beyond this point. Another concept – one that was both modeled and tested – was a train-mobile scheme that would take advantage of the existing railroad network in the Midwest and Western United States. Up to 5 squadrons and 50 trains were planned, each with five transporter-erector-launcher cars and supporting cars. Only 90 missiles (18 trains) were ever actually funded, but these were canceled during the Kennedy administration.

In the end Minuteman was deployed in underground hardened missile silos scattered throughout the midwestern United States. Some 450 Minuteman I launchers would be authorized in the Eisenhower administration. All in all, the small size and solid propellant of the Minuteman, as well as its self-contained inertial guidance system, did provide advantages other than mobility to Air Force planners. The Minuteman could be launched at nearly a moment's notice compared to the 30 minutes it took for Atlas. Also fewer men were required for actually launching the Minuteman missile. Consequently, when all 1000 launchers became operational, at most only 200 launch officers would be required to fire all the missiles. Minuteman I carried a 1.3 megaton warhead with an accuracy of about 1.0-NM CEP. Some 800 Minuteman I missiles were emplaced in silos.

Pershing

The Pershing I theater-range ballistic missile was developed by the Martin Corporation beginning in 1957 and was to be a replacement for the unwieldy Redstone theater ballistic missile. Flight testing began in 1960 and by 1962 Pershing I was in Army service. During 1964 it was first deployed overseas with the US Seventh Army in West Germany. Pershing is a solid-propellant two-stage missile that contains an inertial guidance system producing a 0.50-NM CEP with an insertible pre-launch targeting package. The MGM-31A Pershing I weighs 10,141 lb and the two stage rocket propulsion by the Thiokol company gives ranges up to 460 miles. Inevitably Martin Marietta, the prime contractor, had to build a complex system which traveled on tracked vehicles. By 1967 this was refined in Pershing IA to a set of wheeled vehicles, which are transportable in C-130 Hercules aircraft and carry the missile already assembled with its 400-kiloton warhead. Further development by 1976 enabled one commander to fire up to three missiles after only a brief delay following arrival at a site which had not been previously surveyed. One

Top: A Martin Marietta Pershing 1a tactical battlefield missile on its transporter.
Left: Test firing of a Pershing 1. Subsequently Pershing 2 was developed with much greater accuracy, enabling a smaller warhead to be used.
Above: Two dummy XGAM-87A Skybolt ALBMs hung under the wing of a B-52F trials aircraft.
Right: Only 54 Titan IIs entered service with the USAF. This silo is furnished with work platforms at every level all round the missile.

hundred and eight Pershing I missile launchers in three battalions are assigned to the 56th Field Artillery Brigade deployed in West Germany. Another 72 Pershing I missile launchers were delivered to the West German Luftwaffe, under a dual control arrangement, for its use. Pershing I, because of its yield and range (not to mention its easier-to-handle solid propellant) is a formidable theater strike weapon capable of attacks against Warsaw Pact airbases and casernes.

Skybolt

The Skybolt was an air-launched ballistic missile built by Douglas Aircraft beginning in 1959 and flight tested in 1961. Skybolt was canceled in 1962 because it was decided that the concept had the disadvantages of both a bomber and a missile. Additional Minuteman IIs (150) were purchased in its place. Skybolt had a two-stage solid-propellant motor, inertial guidance and a range of 850 NMs. Its warhead may have been similar to the 1.3-megaton warhead that then existed on Minuteman I. Before it was canceled Skybolt was scheduled to be installed on 102 USAF B-52 bombers (4 each or 408 Skybolts) and perhaps 120 British Vulcan bombers (2 each or 240 Skybolts). Skybolt was to be the offensive punch of SAC's and RAF Bomber Command's bomber force during the 1960s. As a stand-off air-launched ballistic missile it was to be mated with two other

bombers on the drawing board – the WS-110A (later the B-70) chemical-powered bomber and the WS-1252 nuclear-powered bomber. Neither was produced. Skybolt was a joint project with the British (for which the British Blue Streak IRBM was canceled), so when Skybolt development ended and Thor's future was uncertain, the British (to maintain the agreed-upon nuclear deterrent capability) opted for Polaris A-3 missiles for a yet-to-be-built submarine force. Skybolt was to be an all-azimuth ballistic missile either to strike Soviet targets directly or cut corridors for bombers to deliver higher megaton-yield payloads.

Titan II

Titan II was the last missile design pursued during the Eisenhower administration and the last large ICBM built by the United States. Titan II, built by Martin, was intended to be a second-generation heavy ICBM intended to replace Atlas and Titan I. Titan II began development in 1958, was flight tested in 1961 and entered service in 1963. Titan had storable liquid propellant, giving it a reaction time of about one minute, and self-contained inertial guidance with no radio command necessary. Guidance improvements were made in the late 1970s. Titan II's missile yield was 7.4 megatons. Fifty-four Titan IIs were deployed in six squadrons of nine missiles each, although initially the force was to have 108 launchers. The Titan II project was a response to both Soviet ICBM testing and the slowness and vulnerability of liquid-fueled, semiprotected Atlas and Titan I missile systems. Titan II's deployment into hardened underground missile silos ensured its future with the SAC missile force, where it has remained in service for twenty years fulfilling a variety of specialized targeting requirements that would take advantage of its hard-to-reach bases, high yield and exceptionally long range. Following an accident in 1980 the number of Titan IIs in service was reduced by one to 53.

The Doctrine of Flexible Response

As the Eisenhower administration left office, the Kennedy regime inherited the reins of power and the responsibility for maintaining deterrence for the US as well as its allies. But numerous strategic missile programs had yet to bear fruit. In addition, many other strategic programs already in place were, for a variety of technical and strategic reasons, obsolete before they were even deployed.

Furthermore, the missile gap that was feared between the USSR and the US failed to materialize in the intercontinental field, although large-scale deployments of Soviet regional missiles did occur in this period. Awareness of the Soviet Union's military infrastructure was also unfolding as a result of active reconnaissance of the Soviet Union by aircraft such as the U-2 and, soon, first-generation reconnaissance satellites. This would further expand US targeting knowledge and support a transition in US strategy. As a result, by the end of the 1960s US strategic missile forces would not only reach full maturity in survivability and capability, but would also provide balance and a complement to the nation's strategic bomber force.

Much of the Kennedy administration's impact in the national security arena took the form of new directions in US defense policy and weapons programs resulting from dissatisfaction with the declared policy of massive retaliation developed during the Eisenhower administration.

In the Kennedy administration's message to Congress on the defense budget given in early 1961, it was asserted that a policy of flexible response would steer the administration's new national security policy. Soviet or Chinese attacks or actions in the world would be offset in a deliberate, expedient and effective manner rather than through the spasm response expected in the Eisenhower years. In addition to a new declaratory policy, there were also important criteria that would have to be met in future weapons-systems procurement choices involving force survivability and responsiveness.

By 1961 US strategic forces were undergoing a gradual evolution into a modern strategic force. The new weapons systems developed after 1954 would soon be entering the US strategic inventory in great numbers. A significant portion of this force would also be sea based and operated by the Navy rather than the Air Force. This, as can be imagined, might have caused some conflicts in mission and targeting assignments. Indeed, at the time the Navy had already suggested that SAC (bombers or land-based missiles) capabilities could only provide surplus force above what the Navy SLBM/SSBN force would be capable of in a short time.

Partly in response to both the political infighting developing and the one-strike nature of SAC nuclear weapons employment policy as it evolved in the Eisenhower years, the Defense Department issued in 1960 the Gates (Secretary of Defense) Report which led to the establishment of a Joint Strategic Target Planning Staff. From then on there would be annual planning for the single integrated operations plan (SIOP), the US war plan which directed the placement of forces and targeting. The SIOP was an operational plan that laid the foundation for a coherent US nuclear war planning and fighting strategy. In the last year of the Eisenhower administration this plan, with the forces in place at that time, called for a single large-scale attack at the onset of war on a variety of Soviet and Chinese nuclear and conventional military targets as well as the Soviet Union's industrial base.

The first year of the SIOP revision was also the first year of the Kennedy administration. One result of this was greater emphasis on flexibility in the war plan. Consequently, the first targeting revision distinguished among the three general targeting categories above in terms of both priority and timing. In effect, Soviet and Chinese military forces and urban industrial targets would not necessarily be struck with nuclear weapons at the same time. More so, conventional and nuclear military targets might also be split into separate targeting missions. Much of the operational targeting philosophy of the new administration from the early 1960s onward reflected this discriminating approach to war planning and war fighting.

The Kennedy administration and the flexible re-

Launch of a Titan II ICBM from Vandenberg. This missile, now being withdrawn, was the only one in the West to begin to approach the range and throw-weight of Soviet ICBMs.

Missile	Designer	Year Design Begun	First Flight Test	Propulsion System	Guidance	Warhead	Range	Year Operations Began	Weight at Liftoff	Basing Mode	Number Deployed
Nike Ajax NIM-3	Western Electric	1945	–	Solid fuel	Command	High explosive	20 NM	1953	2 tons	Rail	–
Nike Hercules MIM-14	Western Electric	1951	–	Solid fuel	Command	HE/ Nuclear	75 NM	1958	5 tons	Shelters	–
Bomarc	Boeing	1945	1962	Ramjet Active	Command	Nuclear	380 NM	1960	8 tons	Shelters	–
WIZARD	Convair	1954	–	Solid fuel	Command	–	–	–	–	–	–
Nike Zeus	Bell	1954	1959	Solid fuel	Command	Nuclear MT yield	400 NM	–	20 tons	–	–
Nike X System Spartan	Bell	1963	1968	Solid fuel	Command	Nuclear 4.0 MT	400 NM	1975	14 tons	Silo	–
Nike X System Sprint	Martin	1963	1965	Solid fuel	Command	Nuclear 10 KT	20 NM	1975	4 tons	Silo	–

sponse doctrine had other issues to resolve. While various weapons policies would satisfy many of the new targeting requirements, strategic problems, like force vulnerability to enemy attack, also had to be addressed.

Various measures to offset force vulnerability in the near time by the Kennedy administration included in the early 1960s an increase in bomber alert rates, better protection for political and military command and numerous retirements of missiles that were vulnerable to surprise attack although they had only recently entered service. In early 1961 the single Snark squadron in Maine was retired. Snark, and later Skybolt, were believed by the Kennedy administration to have the disadvantages of both the bomber (slow and vulnerable to air defenses) and the missile (no recall and small payload). Regulus was also to be phased out of the Navy because of its prelaunch vulnerability as well as its susceptibility to air defense. In addition, prelaunch vulnerability canceled a modification to the Long Beach cruiser due to carry eight Polaris missiles. Furthermore, Thor, Jupiter, Atlas and Titan all had liquid propellants and slow response to the point that they could be preempted in a surprise attack. Most were also in above-ground launch complexes or lightly protected. As a result, a premium in war would be on launching the missiles as soon as possible. The eventual phasing out of Jupiter and Thor and their replacement by US submarines indicated a drift away from the Eisenhower administration's concept of regional based deterrence to one that was operated solely by the United States.

The Missile Gap
The Kennedy administration entered office in part because of the 'missile gap' issue in which the Soviets were assumed to be in the lead with ICBMs compared to the US. While some months passed before space-borne reconnaissance systems reported that Soviet ICBM activity was light (while MRBM and IRBM activity was heavy), the Kennedy administration still had to decide

on the best course for future US strategic forces, whether as a political instrument providing an umbrella for NATO, or as a hedge against possible Soviet developments.

Under the Kennedy administration, future weapon choices would require a new framework wherein both Air Force and Navy missiles would be evaluated in a coordinated force-planning effort rather than separately as was the case in the late 1950s.

The new framework attempted to establish objectives for the various force programs under development, regardless of Soviet progress, not so much for offensive and defensive missions, as in the Eisenhower Administration, but for assured-destruction and damage-limitation requirements.

Assured destruction was the deterrence of a nuclear attack on the US or its allies by having in place a military force capable of extreme destruction upon the Soviet Union or the Chinese even after a surprise attack. Damage limitation, on the other hand, stressed limiting damage to the population and industrial bases of the US through active and civil defenses.

Out of these broad new requirement concepts came the US strategic force posture – Minuteman, Poseidon and Trident – (minus the active defenses as a result of SALT) that are in place today. But what was deemed adequate did not match the original Air Force and Navy proposals. Indeed, the Air Force argued successively for as many as 10,000, 2500 and 1200 Minuteman launchers. In the end they received 1000. Similarly, the Navy argued for 45 SSBNS and 720 launchers but on their own timetable. At its peak, the Navy settled for 41 SSBNs with 656 launchers and instructions to make construction a national priority. In sum, there was no one formula for determining future US force levels. Under the Kennedy administration, however, excess force was deemed not to be worth the added expense for the low assured-destruction or damage-limiting value it would have had beyond forces already planned.

Missile Development 1961-69

As the US missile force modernized along various guidelines, it was clear that these were more powerful and reliable weapons than the first-generation weapons already discussed. They were also more reliable and survivable in the face of Soviet improvements and incorporated new technological advances such as MIRVed warheads. These missiles, deployed in silos, at sea and on bombers, would be the backbone of US strategic forces through the 1980s.

The Poseidon C-3 sea-launched ballistic missile was a larger more accurate missile designed by Lockheed to replace the Navy's Polaris missiles beginning in 1961. It was tested in 1968 and became operational in 1971. Poseidon is a solid-propellant missile with inertial guidance. Its most innovative feature was its multiple independently targeted re-entry vehicle (MIRV), which allowed in this case up to 14 warheads of 40 kilotons each to be carried by one missile. MIRVing a missile allowed greater efficiency in target coverage. Some 496 launchers on 31 submarines were deployed.

The Minuteman II ICBM designed by Boeing as a replacement for Minuteman I began development in the beginning of 1962. It was test flown in 1964 and deployed in 1966. Minuteman II is a second-generation three-stage solid-propellant missile with inertial guidance and a 1.1-megaton warhead. Minuteman II would also have decoys in its warhead to assist warhead penetration in the event the Soviets deployed anti-ballistic missiles, as was expected. Furthermore, each Minuteman II could store information on eight targets prior to launch rather than on only one, as in Minuteman I. Five hundred Minuteman IIs were deployed in

Top: Primary weapon of the limited-range FB-111A swing-wing bomber of SAC is the small SRAM (short-range attack missile), four of which can be carried on wing pylons.
Right above: Launch of an LGM-30G Minuteman III from Vandenberg AFB on 19 April 1979.
Right below: The three warheads of a Minuteman III ICBM, with their protective nosecone at the right.
Above: Salvo firing of Minuteman II ICBMs Nos 85 and 86 from training silos at Vandenberg.

300-psi silos, replacing some 300 Minuteman Is and giving the US some 1000 Minuteman missiles in service by 1967.

The air-launched Short Range Attack Missile (SRAM) began development in 1963 by Boeing, was tested in 1969 and entered service in 1972. SRAM is a solid-propellant Mach-3.0 inertially guided missile with a 170-kiloton warhead. Fifteen hundred SRAMs were delivered to SAC and can be fired from B-52s or FB-111s. B-52s could carry as many as 20, while FB-111s could carry six. In typical operational missions B-52s probably carried four SRAMs and four 5.0-megaton bombs, while FB-111s carried four SRAMs and two 1.3-megaton bombs. SRAM was developed to assist bomber-penetration routes by directly attacking enemy early-warning and air-defense sites, thus freeing ICBMs for other deep-strike missions. SRAM signified the continuing importance of missile development for assisting manned bomber operations.

The Minuteman III ICBM was also designed by Boeing. It began development in 1964 and entered service in 1970. It is a third-generation three-stage solid-propellant rocket with inertial guidance. Importantly, it was equipped with a MIRV'd system carrying up to three 170-kiloton warheads, initially to within a 0.25-NM CEP from the target; it could also be retargeted quickly because of the Command Buffer System. This system allowed Minuteman III to be retargeted prior to launch in 36 minutes versus 20 hours with Minuteman II. Five hundred and fifty Minuteman III missiles were put into 2000-psi missile silos in the midwest, replacing some 50 Minuteman IIs. The introduction of Minuteman III equipped with multiple warheads, together with Poseidon and SRAM, increased greatly the number of weapons available to the US for retaliation even in the event of a surprise attack.

Deferred Missile Systems

The mobile medium-range ballistic missile program initiated in 1962 was to be a 2600-NM, two-stage solid-propellant missile, a land-based replacement for Thor and Jupiter. The program was not pursued after 1964.

The WS-120A was an advanced ICBM that may have been intended to be Minuteman IV. WS-120A was to be a three-stage solid-propellant missile with inertial guidance and a 0.15-NM CEP. It would have incorporated a high-performance solid-propellant rocket motor and advanced quasi-fourth-stage Post Boost control vehicle. The missile began development in 1963. Testing was scheduled for 1971 and 500 were scheduled for introduction into service between 1973 and 1975. The MIRVed configuration of WS-120A was uncertain, but would have had a high probability of destroying very hard missile silos in the Soviet Union with one or two warheads. WS-120A was to be silo based in hard rock silos in the midwest. WS-120A was deferred, but it laid the foundation for future US ICBM development.

Strategic Sufficiency to Countervailing Strategy

When the Nixon administration took office in 1969, a modification of earlier strategic concepts emerged. The administration concluded that strategic nuclear missile superiority was impossible to achieve. Consequently, the US would tailor its strategic programs to ensure that the Soviets did not develop an advantage in any field of strategic weapons. If advantages were perceived, the US would expand its programs accordingly. As a result, strategic sufficiency would attempt to protect both US and Soviet deterrent forces. In this context the Nixon administration, as will be seen, revised the Sentinel ABM program to one called Safeguard, having as its mission the defense of Minuteman missiles. The strategic landmark for the Nixon administration was, of course, the SALT negotiations which began in 1969 in an effort to end the arms race.

The last review of US nuclear targeting strategy had been completed in 1960. Targeting between that point and 1973 had been a steady progression of adaptation, driven not by any institutional process but by changes in the military environment. By the early 1970s concern arose as to the procedures involved. A number of studies for the Nixon administration were completed by the Defense Department, and in 1974 the DoD implemented the Nuclear Weapons Employment Policy.

This employment policy had three major parts. First, it stressed that the targeting of Soviet military forces would continue. Second, it emphasized that escalation would be flexible in how it evolved, and, finally, it stipulated that some targets would not be struck at all while others would be held for bargaining purposes. In any event, force requirements would stress both highly effective and survivable forces, both for initial strikes and as a strategic reserve. All US strategic forces would have as a basic requirement the destruction of 70 percent of Soviet industry that could be used in a recovery following World War III. Within this framework a new SIOP guided the wartime employment of US strategic forces.

Countervailing Strategy

When the Carter administration assumed power in 1977, no change in targeting policy took place. In fact, through Presidential Directive 18 (PD-18), the Carter administration codified all aspects of existing strategic policy as well as studying new employment policy, ICBM modernization issues and the importance of nuclear reserve forces in a war. The results of these studies culminated during 1980 in PD-59. Known also as a countervailing strategy, this approach was conceived to enhance the deterrence of nuclear war by assuring that whatever the level of aggression contemplated, no potential adversary could ever conclude that victory, in any meaningful sense of the word, would be attainable or worth the costs that would be incurred.

Consequently, the current war plan includes four general categories of targets. This version of the SIOP has been under institutional development since 1974. The execution of the SIOP can be accomplished in one of four ways, and within that framework there can be targets that are not struck: national command and control centers or urban areas, as well as potentially hostile countries like China which are an option but which would not automatically be attacked.

There are also special sets of targets to be struck within these options in the case of either preemptive attacks or launch-under-attack scenarios. Finally, there is a need for continued target location updating for later strikes. All are the blueprint for World War III, but have as their objective the deterrence of a nuclear attack on the US or US forces or allies, or, failing that, the ability to inflict an appropriate retaliatory response on the USSR or aggressor.

Nuclear Weapon Employment Options

I **Major Attack Option**
The complete destruction of an enemy by US strategic forces

II **Selected Attack Option**
Destruction of enemy military forces in the homeland prior to their use elsewhere.

III **Limited Nuclear Option**
Selective destruction of fixed enemy-homeland military or industrial targets

IV **Regional Nuclear Option**
Destruction of enemy military forces outside the homeland from areas outside the continental US

SIOP Target List

1. Soviet nuclear forces:
ICBMs and IRBMs, together with their launch facilities (LFs) and launch command centers (LCCs).
Nuclear weapons storage sites.
Airfields supporting nuclear-capable aircraft.
Nuclear, missile-firing submarine (SSBN) bases.

2. Conventional military forces:
Casernes.
Supply depots.
Marshaling points.
Conventional air fields.
Ammunition storage facilities.
Tank and vehicle storage yards.

3. Military and political leadership:
Command posts.
Key communications facilities.

4. Economic and industrial targets:
a. War-supporting industry
  Ammunition factories.
  Tank and armored personnel carrier factories.
  Petroleum refineries.
  Railway yards and repair facilities.
b. Industry that contributes to economic recovery
  Coal.
  Basic steel.
  Basic aluminum.
  Cement.
  Electric power.

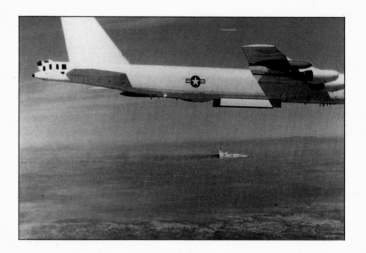

Above: The first test drop of an AGM-86A from the B-52G trials aircraft on 5 March 1976.
Below: Assembly of McDonnell Douglas guidance systems for four versions of AGM/BGM-109 (Tomahawk) cruise missile.
Right above: Exploded diagram of the Boeing AGM-86A, the original (short-body, short-range) ALCM.
Right: Comparative left-side photographs of an AGM-86B ALCM (upper) and AGM-109A Tomahawk (lower).

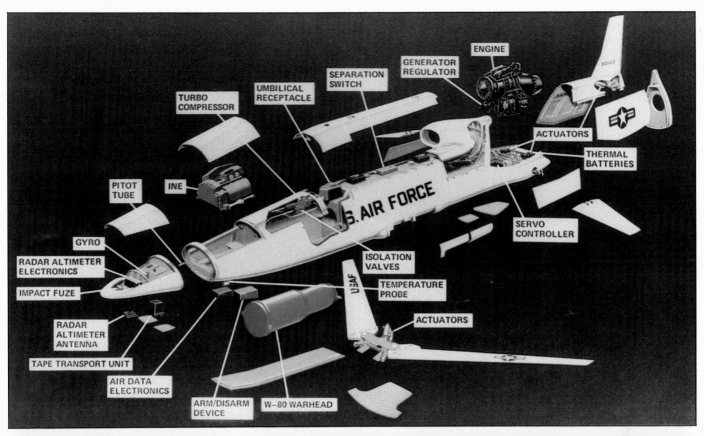

TURBO COMPRESSOR

UMBILICAL RECEPTACLE

SEPARATION SWITCH

GENERATOR REGULATOR

ENGINE

ACTUATORS

THERMAL BATTERIES

PITOT TUBE

INE

SERVO CONTROLLER

GYRO

RADAR ALTIMETER ELECTRONICS

IMPACT FUZE

ISOLATION VALVES

TEMPERATURE PROBE

RADAR ALTIMETER ANTENNA

ACTUATORS

TAPE TRANSPORT UNIT

AIR DATA ELECTRONICS

ARM/DISARM DEVICE

W–80 WARHEAD

U.S. AIR FORCE   ALCM

U.S. AIR FORCE   AGM-109

Missiles of the Contemporary Period

The Air Launched Cruise Missile (ALCM) began development in 1970 based on the SCAD (Subsonic Cruise Armed Decoy) Quail replacement project. ALCM was test flown in 1976 and entered service with SAC in 1982. ALCM is a subsonic turbofan-powered aerodynamic missile (whose technology is beyond that of the World War II and post-war cruise missiles discussed earlier) with both inertial guidance and a terrain contour-matching system (TERCOM). This system periodically updates the missile track to the target by comparing actual terrain with prestored mapping data in the missile computer. This last feature gives the missile its high accuracy. A total of 3400 ALCMs, each with a 250-kiloton warhead, is scheduled to be produced over the life of the program. B-52s can carry as many as 20 ALCMs (12 on wing pylons and eight in the bomb bay). With a 1500-NM range, the ALCM is an excellent standoff weapon for the B-52, allowing the airplane to stay out of Soviet air-defense range but still strike primary targets.

The Trident submarine, missile and warhead programs are designed to provide the US with a new class of fleet ballistic missile submarines and associated submarine-launched ballistic missiles to augment and eventually replace the present Poseidon force.

This new missile and submarine resulted from a series of studies (called Strat-X) that began in the late 1960s and culminated in 1971 in the Undersea Long Range Missile System (ULMS). The USS *Ohio* SSBN was launched in 1980, displaces 18,000 tons and can launch 24 Trident missiles. Designed by Lockheed starting in 1971, the Trident I missile was flight tested in 1977 and became operational in 1979. Trident I is a three-stage solid-propellant SLBM with a range of 4000 NMs. The first two stages of the Trident I missile are Poseidon, so in initial design configuration it was called Extended-Range Poseidon (EXPO). Guidance is inertial and incorporates a stellar sensor that allows high accuracy compared even to shorter-range missile systems. The Trident I can carry eight MIRVed warheads of 100 kilotons each. Later versions of the Trident I may carry Mk 500 maneuverable warheads

The first Trident I replaced Poseidon SLBMs on board the most modern *Lafayette* Class SSBNs, 12 of which are equipped with Trident Is (192 Missiles), as will be the new *Ohio* Class SSBNs. How many will be operational on the 15 projected *Ohio* Class will be based on military decisions about Trident II as a replacement for Trident I. One hundred other Trident I SLBMs have been ordered by Britain.

The MX (Peacekeeper) ICBM began as a SAC requirement for an advanced ICBM in 1971. Air Force development in 1973 focused on the technology and design concepts required for a large, survivable, highly accurate MIRVed ICBM. Boeing began advanced development in 1976 and full-scale development in 1979. The first flight test for Peacekeeper is scheduled for 1983, with the weapon due for entry into service in 1986. Peacekeeper is a three-stage solid-propellant missile with advanced inertial guidance. Peacekeeper has a MIRVed post-boost reentry vehicle that can carry ten warheads – each of 335 kilotons. The system may have an accuracy of less than 0.10-NM CEP. Peacekeeper is scheduled to be a partial replacement of the Minuteman III force with as many as 100 planned (not including an additional 75 for spares). Several basing concepts for Peacekeeper have been proposed, including Minute-

Missile	Designer	Year Design Began	First Flight Test	Propulsion System	Guidance	Warhead	Range	Year Operations Began	Weight at Liftoff	Basing Mode	Number Deployed
XB63; GAM-63 Rascal ASM	Bell	1946	1953	Rocket	Radio Inertial 0.25 NM CEP	One	75 NM	1957	6.5 tons	B-47	
GAM-72; ADM-20 Quail ASM	McDonnell	1953	1958	Rocket	Pre-planned Path	ECM	250 NM in 30 min	1960	0.5 ton	B-52	492
GAM-77; AGM-28 Hound Dog ASM (formerly WS-131)	North American	1955	1959	Turbojet	Inertial	One 4.0 MT	700 NM	1961 to 1976	5 tons	B-52	593
AGM-69 SRAM ASM (formerly WS-140A)	Boeing	1963	1969	Solid fuel	Inertial 0.25 NM CEP	One 170 KT	100 NM	1972	1 ton	B-52 FB-111	
AGM-86B ALCM ASM	Boeing	1970	1976	Turbofan	Inertial 0.05 NM CEP	One 250 KT	1500 NM	1982	1 ton	B-52	
XSM-64 Navaho Cruise Missile (formerly WS-104)	North American	1947	1958	Liquid fuel	Astro Inertial	One	6250 NM at Mach 3.0	–	145 tons	Above ground Pad	–
XSM-73 Goose Cruise Missile	Fairchild	1951	1958	Liquid fuel	Inertial	ECM Jammers	–	–	2.5 tons	Above-ground Pad	–
GAM-87 Skybolt ASM (formerly WS-138A)	Douglas	1958	1961	Solid fuel	Inertial	One	1150 NM	–	5.5 tons	B-52 Vulcan	
TM-61; MGM-1C Matador	Martin	1947	1953	Turbojet	Radio Command	One 400 KT	650 NM	1955	6.0 tons	Pad	144
TM-76 Mace	Martin	1953	1955	Turbojet	Radio and Radar	One 4 MT	1100 NM	1959	9.0 tons	Shelter	172
SSM-A-14 Redstone	Redstone Arsenal	1950	1953	Liquid fuel	Inertial	One 10 KT	215 NM	1958	30 tons	Mobile	16
MGM-31 Pershing	Martin	1957	1960	Solid fuel	Inertial	One 60-400 KT	400 NM	1962	5 tons	Mobile	108
Pershing II	Martin	1974	1982	Solid fuel	Terminal	One 1-10 KT	1000 NM	1983	5 tons	Mobile	108 planned
Tomahawk GLCM	General Dynamics	1977	1980	Turbofan	TERCOM	One 200 KT	1350 NM	1983	1.5 tons	Mobile	464 planned

SSBN/SLBM Schedule
Cumulative Forces After Initial Sea Trials

	78	79	80	81	82	83	84	85	86	87	88
TRIDENT SSBNs	0	0	0	1	3	4	6	7	9	10	11
with TRIDENT I SLBMs	0	0	0	24	72	96	144	168	216	240	264
POSEIDON SSBNs	0	0	5	11	12	12	12	12	12	12	12
with TRIDENT I SLBMs	0	0	80	128	192	192	192	192	192	192	192
POSEIDON SSBNs	31	31	26	20	19	19	19	19	19	19	19
with C-3 SLBMs	496	496	416	368	304	304	304	304	304	304	304
POLARIS SSBNs	10	10	5	0	0	0	0	0	0	0	0
with A-3 SLBMs	160	160	80	0	0	0	0	0	0	0	0

Submarine Basing – 1988
Charleston: 10 *Lafayette* Class (160 Poseidon C-3s)
King's Bay: 12 *Lafayette* Class (192) Trident Is)

Bangor: 11 *Ohio* Class (264 Trident Is)
Holy Loch, Scotland: 9 *Lafayette* Class (144 Poseidon C-3s)

The first launch of a C-4 Trident SLBM took place from a land
pad at Cape Canaveral on 18 January 1977. Note the spike added
to the nose for aerodynamic purposes.

Far left: Sub-surface launch of a Trident C-4. This program was severely hit by delays and cost-escalation, as were the submarines being built to carry it.

Top: Some 20,000 guests attended the launch by Electric Boat of SSBN-726 USS *Ohio*, lead ship of the great Trident-carrying class, at Groton on 4 April 1979. At right is No 727 *Michigan* and the keel ring of 729 *Georgia*.

Left: Technicians working in an SLBM launch tube aboard a US Navy SSBN.

Above: USS *Ohio* (SSBN-726), first of the giant new Trident submarines, at sea in September 1981.

man and Titan silo basing, air mobile systems, a race-track pattern with multiple protective shelters (200 missiles scattered among 4600 shelters using both concealment and mobility for survivability) and currently a concept labeled Dense Pack. Dense Pack would have 100 missiles located in silos in a one-by-fourteen-mile rectangle. Each of the silos in turn would be reinforced to withstand 10,000 psi of blast. As a result the Peacekeeper force is believed capable of withstanding total preemption because of hardness of the individual silos and the 'fratricide' effect of nearby explosions on other incoming reentry vehicles.

Although not a new missile program, the Mk-12A reentry vehicle is being acquired to replace the Mk-12 RV on 300 of the 550 Minuteman III missiles currently deployed. Initial engineering work began on the Mk-12A in 1974 and initial operational capability was achieved in 1980. Mk-12A has a yield of 335 kilotons compared to the 170 kilotons on Minuteman. Current plans do not call for the Mk-12A to be deployed on the remaining 250 Minuteman III missiles. One hundred of these missiles are in any case likely to be replaced by Peacekeeper. The Mk-12A RV was designed to be employed against a variety of targets, but has been increasingly planned for employment against a growing Soviet hardened-target system, where its combination of yield and accuracy could be used to military advantage. Although the Mk-12 in its current configuration is effective to some degree against hard targets, improved accuracies may accrue as a result of guidance improvement programs. The higher yield of the Mk-12A warhead would increase this capability. Combined, Minuteman and Peacekeeper will have 1900 hard-target weapons.

The BGM-109A Tomahawk sea-launched cruise missile (SLCM) began full-scale development by General Dynamics in 1974, was flight tested in 1976 and became operational in 1983 aboard nuclear attack submarines (SSNs). Tomahawk is a subsonic air-breathing turbofan-powered aerodynamic missile. Guidance is inertial and terrain contour-map-matching (TERCOM) provides extremely high accuracy. Tomahawk is 21 inches in diameter, the size of a torpedo, and has a nuclear warhead of 200 kilotons. The strategic attack version of the Tomahawk is now deployed aboard SSNs in place of torpedoes. Original plans called for four SLCMs to be placed in each of the 16 launch tubes on 10 older Polaris submarines. SSNs would typically carry two Tomahawks out of 26 weapon slots that also include torpedoes and Harpoon anti-ship missiles. One modification of the SSN hull has been proposed so that conventional weapon spares would not be compromised. This would be the installation of 12-16 vertical SLCM launchers in the forward part of the submarine. Attack subs would patrol in waters adjacent to the Soviet Union on routine ASW missions. Once war occurred their SLCMs would provide some near-term hard-target kill capability while contributing to a strategic reserve.

The BGM-109B Tomahawk is the ground-launched cruise missile (GLCM) version of the same aerodynamic subsonic missile airframe mentioned above. The

Proposed Number of GLCM Flights in Europe

Country	Flights	Missiles
Britain	10	160
West Germany	7	112
Italy	6	96
Belgium	3	48
Netherlands	3	48

Left: The first completely assembled Peacemaker, then known as the MX, on vibration test at Martin Marietta in 1982.
Above: Testing the separating sequence of the Avco reentry vehicle for the Peacemaker ICBM.

Below: An artist's impression of the CSB (closely spaced basing) 'dense pack' deployment of Peacemaker silos a mere 1,800 ft apart.

General Dynamics GLCM established an identity for itself in 1977 and was test flown in 1980. It is expected to be operational in 1983. This missile has received much attention because of its place in alliance politics after the decision was made in 1979 to reintroduce long-range nuclear forces to NATO. Four hundred and sixty-four GLCM launchers organized into 29 flights are scheduled to deploy in Europe among five countries. GLCMs will be housed at Air Force bases in peacetime and deployed into the field during wartime. Their major mission will be theater nuclear strikes, thereby freeing manned aircraft for conventional missions.

Pershing II is a new missile built by Martin Marietta to replace Pershing I. Pershing II was part of the NATO long-range nuclear modernization program announced in 1979 although advanced development had begun in 1974. The missile was flight tested in 1982 and deployments are to begin in 1983 and end in 1985. Pershing II is a solid-propellant, terminally guided (by a radar area correlation system) two-stage ballistic with a reentry vehicle speed of Mach 8 (Pershing I was credited with Mach 3.9) and is the first Army missile since Jupiter credited with a range greater than 1000 miles.

Today much publicity has attended the repeated test failures of Pershing II. In fact there is no evidence that this new-generation missile is in any way a faulty system, but the noisy anti-nuclear lobby has latched on to European deployment of Pershing II and sought to magnify any of its problems. They appear to have overlooked the central fact that the whole purpose of Pershing II is to replace the devastating warhead of Pershing I with a much smaller warhead. The key to doing so is much greater accuracy. Though much of the rest of the system is unchanged, for obvious reasons of cost, the missile itself has a totally new second stage distinguished by a guided reentry vehicle with four delta control fins and a white nose radome. In the nose is Goodyear area-correlation radar guidance which 'looks' at the terrain on which it is falling, compares it with computer-stored digital information and commands trajectory corrections to bring the two scenes into exact correspondence. Combined with more accurate Singer-Kearfott inertial guidance for the main missile, this should reduce circular error by over 95 percent. In one form Pershing II may be used as a CAAM (Conventional Airfield Attack Missile) with a runway-penetrating conventional warhead while at one time a 90-foot earth penetrator warhead was considered as a way to rout out enemy command bunkers.

The US Army expects to deploy a brigade of Pershing II (three battalions, each with 36 launchers) to West Germany as soon as production can be started. Though unofficial reports credit this missile with ranges exceeding 1000 miles, there is some reason to believe that range will not differ greatly from Pershing IA, though the usual warhead, in the 10-kiloton class, will be lighter, making up for the weight of terminal guidance.

The Trident II or D-5 to be built by Lockheed is intended to take full advantage of the greater volume of room available in the SLBM launchers on board Ohio Class submarines. Full-scale development for Trident II began in 1983, testing could occur by 1987 and fleet operations should begin in 1989. The Trident II will have accuracy improvements over the Trident I, making it a weapon as effective as a modern ICBM with 6000 miles of range. The Trident II will initially have eight warheads of 335 kilotons, until its Mk 600 150-kiloton terminal guidance reentry vehicles are ready in the next decade. Three hundred and sixty could eventually be deployed aboard Ohio Class SSBNs, giving this force 2880 hard-target weapons.

US Air Defense Missiles
In the mid-1950s US military planners believed that the Soviet Union would build up its bomber force not only to attack targets in Europe and Asia, but to strike the United States and Canada. A number of reasons account for this, including a 'bomber bias' in the US Air Force and the lack of appreciation of ballistic-missile research and development efforts in the Soviet Union.

It was expected that the Soviet Union would deploy up to 500 Bear and Bison bombers for these missions. While U-2 missions would reduce this estimate through the late 1950s (to 375, then down to 300), the air defenses of the United States and Canada became a priority mission in the event of a future World War. While the bomber threat never materialized (some 200 are now in service, adapted for a variety of missions, including anti-carrier attacks), the creation of the Soviet Strategic Rocket Forces resulted in the development and deployment of a number of US strategic air-defense programs. In 1961 this force included 1704 interceptors (75 squadrons) and 4249 rockets and missiles (216 squadrons and batteries). As the threat shifted to ICBMs, the air-defense forces diminished in size.

The rockets and missiles assigned to the North American Air Defense Command included three types: the Army's Nike Ajax and Nike Hercules and the Air Force's Bomarc. The result of service competitions, like Jupiter and Thor, they were eventually both funded and deployed.

Development of US Ballistic Missiles, by service, 1946-80

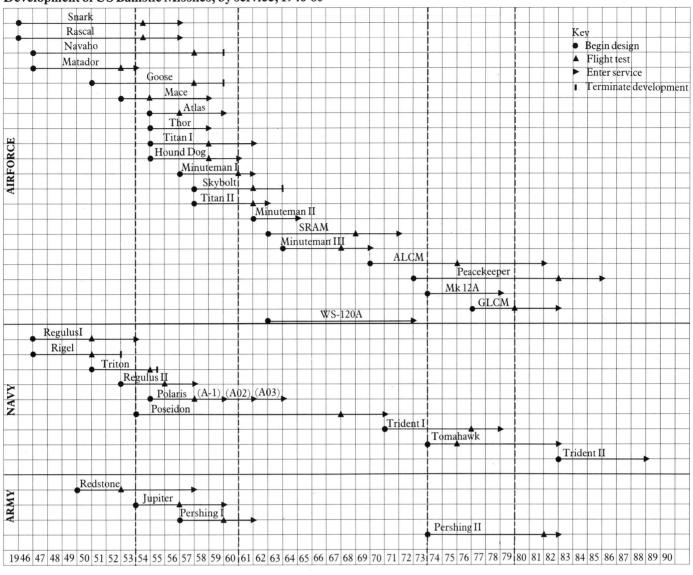

Key
- ● Begin design
- ▲ Flight test
- ► Enter service
- ▮ Terminate development

AIRFORCE: Snark, Rascal, Navaho, Matador, Goose, Mace, Atlas, Thor, Titan I, Hound Dog, Minuteman I, Skybolt, Titan II, Minuteman II, SRAM, Minuteman III, ALCM, Peacekeeper, Mk 12A, GLCM, WS-120A

NAVY: Regulus I, Rigel, Triton, Regulus II, Polaris (A-1) (A02) (A03), Poseidon, Trident I, Tomahawk, Trident II

ARMY: Redstone, Jupiter, Pershing I, Pershing II

1946 47 48 49 50 51 52 53 54 55 56 57 58 59 60 61 62 63 64 65 66 67 68 69 70 71 72 73 74 75 76 77 78 79 80 81 82 83 84 85 86 87 88 89 90

Below: Nike-Hercules SAM battery north of Fairbanks, Alaska, in September 1962.

Below: Nike-Ajax SAM battery on Staten Island, NY, in the mid-1950s.

The Army's Nike Ajax rocket was developed by the Western Electric and Bell Telephone labs beginning after World War II. Ajax was a Mach-2.3 solid-propellant missile that became operational in 1953. Beginning in 1956 Ajax was tied to a semi-automated air-defense network. Ajax, with a range of 20 NMs, protected urban industrial targets. In 1961 there were 1440 Ajaxes in 69 batteries.

The Army's Nike Hercules solid-propellant rocket was also developed by Western Electric and Bell as a follow-on model to Nike Ajax. Work on this system began in 1951 and Hercules was deployed in 1958. With a larger warhead and greater speed (Mach 3.7), Hercules was at least one order of magnitude more effective than Ajax. Hercules had a range of 75 NMs and could reach aircraft as high as 150,000 feet. Hercules was also intended to protect urban industrial centers. In 1961 2502 Hercules in 139 batteries were deployed.

Bomarc was an Air-Force-operated pilotless aerodynamic interceptor missile developed by Boeing and the University of Michigan. It incorporated solid-propellant booster rockets for launch and ramjet engines for cruising. Bomarc was considered a substitute for manned interceptors. As a result, the Canadian government canceled its Arrow interceptor and bought Bomarc. Development work on Bomarc began in 1945 and testing began in 1952; the missile became oper-

ational in 1960. Unlike Nike, Bomarc was not guided from the ground to intercept its target. Rather, it was guided from the ground while it cruised at 65,000 feet (up to 150,000 feet) until near the target, when an internal target seeker took over. Bomarc was an area defense weapon with a range of 200-380 NMs, depending on the model. Top speed was about Mach 4.0. At its peak seven airbases in the northern United States and two in Canada were equipped with Bomarc in protective shelters. Three hundred and sixty-six Bomarc As and 349 longer-range Bomarc Bs were procured over the life of the system. In 1961 307 Bomarc launchers (eight squadrons) were operational. Final phase-out occurred in 1972.

US ABMs

In 1954 both the Air Force and the Army began looking into the feasibility of intercepting ballistic missiles with other ballistic missiles. The Air Force, in conjunction with Convair and RCA, proposed an anti-ballistic missile (ABM) that would intercept enemy missiles 1000 NMs away. The Air Force ABM, called WIZARD, was an advanced concept that utilized electronic scanning rather than mechanical radars. The project was canceled in light of development work being done by the Army and Bell in the same field.

The first serious effort aimed at the deployment of a US ABM system was the Nike-Zeus program. In

Left above: Artist's impression of the USAF Bomarc long-range SAM.
Left below: Launch of an early Nike Zeus test missile from White Sands Missile Range on 7 January 1960.

Below left: Launch of a Spartan, the long-range missile in the ABM system.
Below: Test firing of a Sprint, the last-ditch defense.

February 1957 the Army awarded Western Electric and Bell Laboratories the prime contractor system responsibility for developing Nike-Zeus. The Army was assigned this new mission as an extension of its previous responsibility for developing and operating land-based surface-to-air missiles. The Nike-Zeus program sought to protect US cities and other targets from Soviet missiles. This program envisioned the deployment of several different radars for detection and control and a Zeus interceptor missile with a range of about 400 miles.

The Nike-Zeus program suffered from numerous technical limitations. Its mechanically steered radars had a limited capacity to handle large numbers of incoming warheads. The interceptor missiles were relatively slow, and therefore had to be launched against their targets before the enemy vehicles had reentered the atmosphere. Consequently, the Nike-Zeus system could not utilize the atmosphere to filter out the debris and penetration aids that concealed the attacking warheads while still in the exoatmosphere.

Technological and political developments led to various reorientations of the Army BMD program in the 1960s. During the mid-decade the Nike-X program (so called for lack of a new name for the Army's ABM) sought to develop a BMD for the protection of a small number of high-value targets like cities. Nike-X incorporated significant technical advances, including more capable phased-array radars that greatly enhanced its target-handling capability and the high-acceleration Sprint missile for interceptions within the atmosphere. Later in the 1960s, a longer-range PAR surveillance radar and a new, long-range interceptor – the Spartan (formerly Nike-X) – replaced their less effective predecessors.

During this period the US BMD program deemphasized the difficult task of defending against a large-scale Soviet missile attack and focused on providing a defense system against a possible Chinese ICBM threat and accidental missile launches. With the September 1967 decision to deploy a BMD system, the program was renamed Sentinel in an attempt to distinguish it from an anti-Soviet BMD system. This reorientation reflected both the US's technological inability to deploy an adequate ABM defense against the growing Soviet missile force and its desire to avoid a new round of strategic arms competition with the USSR. Despite the official reorientation of the US ABM program away from defense against the Soviets, the Sentinel program actually retained the potential for expanding into an anti-Soviet defense system.

Sentinel sought to protect the entire country, including cities, from a light, unsophisticated ICBM attack such as the Chinese could undertake. It envisioned the eventual deployment of six PAR and 17 MSR radars as well as 16 Spartan and nine Sprint interceptor sites, some of which were near major US cities. This would have resulted in the deployment of 1600 Spartans and 900 Sprints. The proposed deployment of ABM sites near cities, however, led to widespread public opposition to Sentinel. This system also had technical weaknesses. Its PARs faced north, and 10 of its MSRs had

Left: Launch of two Spartan ABMs from Meck Island (Kwajalein group), one of which intercepted an ICBM re-entry vehicle fired from California.
Above: Sprint ABM at White Sands in January 1966.

only one or two faces, seriously limiting their coverage. Only nine of the installations, including only four of eight coastal installations, were defended by Sprints. They were thus vulnerable to SLBM attack.

For these and other reasons, in March 1968 President Nixon redirected the ABM program away from population defense and toward a phased deployment of defense of the US deterrent forces, especially the Minuteman ICBMs. This new ABM system, called Safeguard, inherited Sentinel components. It envisioned 12 sites, all within the continental United States. This would have netted a total of 360 Spartans and 840 Sprints. Each would have 360-degree MSR coverage, Sprints and Spartans. The system would have fewer Spartans, and many more Sprints, than Sentinel. It would provide area coverage of most of the continental United States. At the same time, it moved the proposed radar and missile sites away from the cities. Consequently, Safeguard retained the mission of defending against a potential Chinese ICBM attack and accidental missile launches.

Although congressional and public opposition raised serious questions about the desirability and effective-

ness of Safeguard, Congress authorized funds for the program. The Phase I plan was to deploy two ABM sites, one to defend Minuteman ICBMs at Grand Forks Air Force Base, North Dakota, and the other at Malmstrom Air Force Base, Montana. Construction of the two initial sites continued until mid-1972, when SALT I ABM Treaty was signed by the President and ratified by Congress. The treaty limited the United States and the Soviet Union to two ABM sites for each nation, one to protect its national capital and the other to defend one ICBM base. Each ABM site was further limited to the deployment of no more than 100 ABM missiles and associated launchers, and a specified number of ABM radars having specified characteristics. Both parties agreed not to develop, test or deploy ABM systems or components that were sea-, air- or mobile land-based. In July 1974 the United States and the USSR signed a protocol that further limited each country to only one ABM site.

The United States chose to complete its deployment at Grand Forks Air Force Base site to defend part of its Minuteman ICBM force. The USSR decided to retain its BMD system around Moscow. Although the site at Grand Forks became fully operational in October 1975, it was deactivated soon after due to the limited defense capability it provided. Furthermore, in 1974 Congress directed the Defense Department to reorient the BMD program to emphasize research and development but not to lead to the testing of a prototype system.

3. EUROPEAN AND CHINESE FORCES

HMS *Renown*; the Royal Navy missile submarines have diving planes on the bow, not on the sail.

Great Britain

During World War II the British had worked with American scientists in developing nuclear weapons, but after the war, in order to maintain their great-power status and deter aggression against their still sizable empire, they began developing their own nuclear weapons. Although a fission (atomic) bomb was tested in 1952 and a fusion (hydrogen) bomb in 1957, the British still had difficulty maintaining parity, technological and otherwise, with the United States and the Soviet Union.

During the 1950s, as in the United States, the British focused their nuclear delivery programs on high-performance subsonic bombers (Valiant, Victor and Vulcan) operated by RAF Bomber Command. Weapons systems for these aircraft included the Blue Boar air-to-surface glide missile to be armed with a kiloton-range warhead and the Blue Steel air-to-surface megaton-range missile. Blue Boar was canceled and Blue Steel became the mainstay of the British bomber force.

Britain also developed its own liquid-fueled IRBM, the Blue Streak. After it was canceled in 1960, a year after Bomber-Command-operated Thors became operational, the British became completely dependent on US strategic submarine programs for their own force developments. Through 1962, for instance, Bomber Command was expecting the Skybolt ALBM to maintain the effectiveness of its bomber forces. When this project was canceled, the Polaris A-3 missile was offered as a replacement to be installed in British-built and Royal-Navy-operated fleet ballistic-missile submarines.

Construction began on the *Resolution* Class in 1964. The first submarine was delivered in 1967; two more were operational in 1968 (*Repulse* and *Renown*) and the last entered service in 1969 (*Revenge*). A fifth submarine was canceled in 1965. Betweeen 1970 and 1981

Summary: Great Britain

Missile	Designer	Year Design Began	First Flight Test	Propulsion System	Guidance	Warhead	Range	Year Operations Began	Weight at Liftoff	Basing Mode	Number Deployed
Blue Streak IRBM	de Havilland	1955	1960	Liquid fuel	Inertial	One 3.0 MT	2500 NM	Canceled in 1960	100 tons	Above-ground Pad	–
Blue Boar	Vickers	1951	1954	Glide	Autopilot T-VI	–	–	Canceled	–	Valiant	–
Blue Steel	Avro	1954	1957	Rocket boost	Inertial with Navigation Update	One 3.0 MT	185 NM	1962	7.5 tons	Valiant Victor Vulcan	60

Far left: HMS *Resolution* leaving Gareloch, Scotland, in February 1977 for a combat patrol.
Left: Royal Navy Polaris A-3 fired from submerged HMS *Resolution*.
Top: HMS *Repulse* entering Portsmouth harbour in November 1976.
Above: A submarine launch of a C-3 Poseidon missile.

the British SSBN force carried out 130 patrols. The original missiles carried by the *Resolution* Class were 16 Polaris A-3, as described above. Beginning in 1973, however, the British began developing an improved warhead for the A-3 missile in a program subsequently known as Chevaline. This included the replacement of the A-3's three 200-kiloton warheads with six 40-kiloton weapons, in effect doubling the number of targets that can be covered by the Royal Navy. The system should be operational by 1983.

Two new missile developments would seem to set the tone of the British-associated missile forces. As already mentioned, 40 GLCM launchers (each with four GLCMs) will be based, beginning in 1983, in England at Greenham Common and Molesworth. They will be nuclear armed and under joint US Air Force/Royal Air Force operational control. Furthermore, in 1981 the British announced they would modernize their national nuclear force of submarines by constructing four new SSBNs, each with 16 launchers, and equipping them with Trident I missiles, some 100 to be purchased from the US for operations and training. The new SSBN force is expected to be operational by the early 1990s and will give the Royal Navy over 500 warheads to target the Warsaw Pact. Some authorities expect that when they are built these submarines will be fitted with the more costly Trident II.

British Missiles
The Blue Streak British IRBM began design work in 1955 under the auspices of de Havilland Aircraft. It was flight tested in 1960 in Australia shortly after it had been canceled by the British Government in favor of Skybolt, an air-launched ballistic missile. Blue Streak was liquid fueled and had inertial guidance; its accuracy may have been half that of the Atlas with a similar 3 MT warhead. Blue Streak would have been stored in a protective silo, although launched and aligned from an above-ground pad. Some 60 Blue Streak IRBMs may have been envisioned by the RAF Bomber Command as a replacement for US-designed Thors and as an offset to Soviet SS-5 IRBM development.

The Blue Boar air-to-surface nuclear-armed glide missile began development in 1951 as a way for bombers to avoid Soviet air defense weapons. It was tested in 1954 and was canceled the same year. Blue Boar was to be carried internally by Royal Air Force bombers and once dropped would deploy pop-out wings to steer it to a target. Guidance was provided by a TV camera located in the missile and beamed back to the mother aircraft. Blue Boar would have had a yield of about 100 KT and could have had a glide range of some 25 NMs, depending on launch altitude.

The Blue Steel air-to-surface rocket began development the same year that Blue Boar was canceled. The missile was tested in 1957 and entered service in 1962, carried by both Victor and Vulcan bombers. At its peak deployment, three Vulcan squadrons and two Victor squadrons (about 60 aircraft and missiles) were equipped with Blue Steel. Blue Steel's range of 175 NM and its varied flight profile allowed Bomber Command aircraft to avoid most existing Soviet air defenses.

Left: Blue Steel stand-off nuclear cruise missiles in an RAF hangar in the early 1960s.
Below: Final assembly of an S 2 SSBS ballistic missile by the Aérospatiale company.
Bottom: Submarine launch of a French M 1, the original MSBS missile.

France

The development of an independent French nuclear force grew out of the political and military requirements of national prestige and uncertainty about the credibility of the US concept of extended deterrence. Although late in starting, French nuclear force development (beginning in 1954) mirrored British development, stressing bombers first and then missiles. France's program, however, was based on its own national effort.

During the late 1950s and early 1960s two French missile programs were started to develop a solid-propellant IRBM and SLBM similar to Polaris. Both became operational in 1971. Tight budgets have kept the force small (the 1st Strategic Missile Group has 18 IRBMs while the French Navy has 5 SSBNs with 80 SLBM launchers), although it is qualitatively advanced. Indeed, the French missile program continues to pursue robust research and development programs which, as seen below, have extended the range of French missiles and increased their yield. Future missile developments are likely to include a new series of SSBNs and a new SLBM, and perhaps later in the decade a land-mobile missile system for the Air Force called SX. This is likely to incorporate three MIRV or MARV warheads.

Summary: France

Missile	Designer	Year Design Begun	First Flight Test	Propulsion System	Guidance	Warhead	Range	Year Operations Begun	Weight at Liftoff	Basing Mode	Number Deployed
M-1	SEREB	1959	1965	Solid fuel	Inertial 0.50 NM CEP	500 KT	1300 NM	1971	20 tons	SSBN	32
M-2	SEREB	1962	1965	Solid fuel	Inertial 0.50 NM CEP	500 KT	1850 NM	1974	22 tons	SSBN	48
M-20	SEREB	1965	1971	Solid fuel	Inertial 0.50 NM CEP	1.2 MT	1850 NM	1977	22 tons	SSBN	80
M-4	SEREB	1976	1980	Solid fuel	Inertial 0.25 NM CEP	MRV 150 KT	2500 NM	1985	40 tons	SSBN	96?
SSBS S-2	SEREB	1959	1961	Solid fuel	Inertial 0.25 NM CEP	One 1.2 MT	1700 NM	1971	35 tons	Silo	18
SSBS	SEREB	1973	1976	Solid fuel	Inertial 0.25 NM CEP	One 1.2 MT	1850 NM	1980	28 tons	Silo	18

Note: All French strategic missile production is now undertaken by a division of the Aérospatiale company.

French Missiles

The S-2 IRBM entered development in 1959. Flight testing began in 1969 and the missile was deployed in 1971. The S-2 is a two-stage solid-propellant missile with inertial guidance. The S-2 has one 150-KT warhead with a 0.25-NM CEP. It is based with the 1st Strategic Missile Group in 18 300-psi silos in the Haute Provence department of France and has a range of 1700 NM. An additional squadron of nine missiles was canceled in 1974. The S-2 was a national effort by the French to offset the prestige associated with the deployment of Thor and Jupiter to Britain, Italy and Turkey in the late 1950s and to provide their own deterrence.

The S-3 IRBM began development in 1973 and was tested in 1976. It entered service in 1980. It is similar to the S-2 except for its more powerful staging motor, which gives it a greater range (1850 NM), and a new 1.2-MT warhead with penetration aids. Reaction time from alert to launch is three minutes. The S-3 has replaced the S-2.

The M-1 SLBM also entered design development in 1959, was tested in 1965 and became operational on *Le Redoutable* SSBN in 1971 and *Le Terrible* SSBN in 1973. M-1 was a two-stage, solid-propellant, inertially

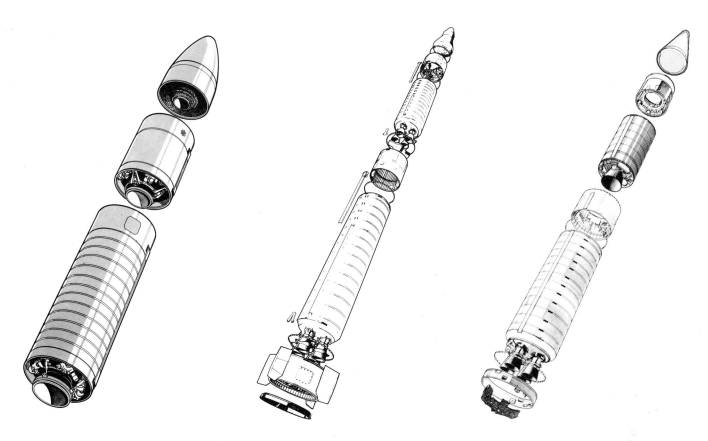

Left: SSBS is the French land-based deterrent loaded into silos on the Plateau d'Albion (Hte Provence) manned by the Air Force.

Above: 'Exploded' drawings of three French ballistic missiles: (left) M 4 MSBS, (center) S 2 SSBS, (right) M 20 MSBS.
Below: Loading a capsule containing an M 20 MSBS into a French SSBN.

guided SLBM with an accuracy of 0.50-NM CEP. It carried a 500-KT warhead of 1300 NMs. The M-2 SLBM replaced the M-1 on the first two French SSBNs and became the first missile deployed with *Le Foudroyant* SSBN in 1974. The M-2 had a more powerful second-stage motor, giving it a range of 1850 NMs. It also carried a 500-KT warhead.

The M-20 was a modification of the M-2. While having the same range and accuracy as the M-2, it carried in addition a 1.2-MT warhead with penetration aids. It was deployed aboard the *L'Indomptable* SSBN in 1977 and *Le Tonnant* in 1980. All of the earlier SSBNs will also be equipped with M-20s.

The M-4 SLBM whose design began in 1976 and was tested in 1980 will be a three-stage solid-propellant inertially guided SLBM with a range of 2500 NM and six MRV warheads of 150 kilotons each. Accuracy is likely to be around 0.25-NM CEP. It will deploy on the sixth French SSBN, *L'Inflexible* (which will itself be an improved submarine) as early as 1985. The M-4 in turn will be backfitted on all but perhaps the earliest SSBN. The continued modernization of the SSBN force has given the French a fairly secure and powerful second-strike force.

Chinese Strategic Forces

Chinese strategic forces were conceived for a number of political and military reasons. These reasons included offsetting US military forces in the Pacific basin, equaling the political prestige its early ally the Soviet Union enjoyed and, as relations collapsed completely, the ability to deter the Soviet Union from attacking China with conventional or nuclear forces. Early Chinese strategic force development, like the Tu-16 Badger bomber force, was based on Chinese initiatives, although the preponderance of training, tooling and fabrication were, no doubt, Soviet supplied.

In 1959 the Soviet Union ended its technical assistance to the Chinese. This had been preceded by the Soviets' withholding selected information about nuclear weapons and Soviet missile developments. Nevertheless, the father of the Chinese missile program, Qian Xvesen, was already working on advanced missiles based on access to the Soviet SS-2 Sibling, itself an improved V-2. From that point long-term plans were developed to test, produce and deploy missiles, each evolutionary in design, that would satisfy Chinese military requirements. Work was further encouraged by the successful test of a Chinese nuclear weapon in 1964.

Chinese strategic missiles today are the product of a well-planned missile force. Chinese missiles offer a credible nuclear retaliatory capability against the Soviet Union throughout Central Asia and the maritime provinces, as well as against US interests in the Pacific. Future strategic programs are likely to include improved force effectiveness and survivability through the introduction of additional silo-based ICBMs, a force of SSBNs and solid propellants for mobile medium-range missiles.

Chinese strategic missiles are operated by the Second Artillery, a special part of the Chinese Army (the Air Force controls the bomber force and the Navy the submarines). The Second Artillery has almost 130 ballistic missiles. The CSS-1, with a 20-kiloton yield and short range, extends the range of Chinese ground forces against Soviet and US tactical targets in Korea. The CSS-2, CSS-3 and CSS-4, with megaton-range warheads and longer range, all have a greater capability against strategic military and urban-industrial targets in the Soviet Union. Chinese missiles are deployed in groups of two or three in the area of the Korean border, south of Peking and south and west of Outer Mongolia.

Summary: China

Missile	Designer	Year Design Began	First Flight Test	Propulsion System	Guidance	Warhead	Range	Year Operations Began	Weight at Liftoff	Basing Mode	Number Deployed
CSS-1	Xvesen	1959	1966	Liquid fuel	Autopilot 1.5 NM CEP	20 KT	600 NM	1970	–	Soft Mobile	50
CSS-2	Xvesen	1962	1969	Liquid fuel	Radio Command	31 MT	500 NM	1972	–	Soft Mobile	75
CSS-3	Xvesen	1963	1970	Liquid fuel	Radio Command 0.75 NM CEP	3 MT	3500 NM	1975	–	Silo	4
CSS-X-4	Xvesen	1973	1980	Storable liquid	Inertial 0.50 NM CEP	4 MT	6500 NM	1982	–	Silo	?
CSS-N-X	China	1975	1982	Solid fuel	Inertial 0.75 NM CEP	One 1 KT	–	–	14 tons	Sub	–

British, French and Chinese Regional Strike Capabilities[1]

	1960	1965	1970	1975	1980	1983
Great Britain						
Bombers	180	80	50[2]	50[2]	48[2]	–
IRBMs	60	–	–	–	–	–
SLBMs	–	–	48	64	64	64
Subtotal	240	80	98	114	112	64
France						
Bombers	–	24	36[3]	36[3]	33[3]	33
IRBMs	–	–	–	18	18	18
SLBMs	–	–	–	48	64	80
Subtotal	–	24	36	102	115	131
China						
Bombers	–	12	18[4]	60[4]	90[4]	
IRBMs	–	–	–	80	122	
SLBMs	–	–	–	–	–	
Subtotal	–	12	18	140	212	
TOTAL	240	116	152	356	439	

[1]British nuclear weapons are operated by the Royal Air Force and Navy; French nuclear forces by the French Air Force, the 1st Strategic Missile Group of the French Air Force and the French Navy; Chinese Forces by the Chinese Air Force and missiles by the 2nd Artillery Command of the Chinese Army.

[2]Vulcan B-2s are configured for low-level attack operations with conventional weapons.
[3]Other Mirage-IVA strike and reconnaissance aircraft in reserve
[4]Mostly Tu-16 Badgers but possibly a few Tu-4 Bulls.

Chinese Missiles

The CSS-1 medium-range ballistic missile began development in 1959, was flight tested in 1966 and became operational in 1970. Some 50 were deployed. The CSS-1 East Wind is a single-stage, slow-reacting, liquid-propellant missile with autopilot guidance, an accuracy of 1.5-NM CEP and a single warhead of 20 KT. The CSS-1 is a further development of the Soviet SS-2 and is based in mountainside complexes. The CSS-1 has a range of 600 NM and, depending on basing, can reach targets in the eastern Soviet Union, some US bases and objectives in Southeast Asia.

The CSS-2 intermediate-range ballistic missile is an improvement of the CSS-1. It began development in 1962, was flight tested in 1969 and became operational in 1972. About 75 CSS-2s are now in service. The CSS-2 is a single-stage, storable-liquid-propellant ballistic missile with radio-command guidance and an accuracy of 0.75 NM. It has a single warhead of 3 MT and a range of 1500 NMs, allowing targets in Soviet Central Asia to be struck. The CSS-2 is also deployed on tracks and in permanent sites in the sides of mountains.

The CSS-3 is a limited-range ICBM made up of CSS-2 components. Development began in 1963, the missile was flight tested in 1970 and a limited deployment of four began in 1975. The CSS-3 is a two-stage, storable-liquid-propellant missile with radio-command guidance and a range of 0.50-NM CEP. Neither the CSS-2 nor the CSS-3 can be salvo launched because of the limitations of their radio-command guidance systems. The CSS-3's warhead is about 3 MT. Importantly, the range of the CSS-3 is 3500 NM, allowing it to strike Soviet targets in European Russia, Moscow excluded. The CSS-3 is based in 300-psi silos.

The CSS-4 is the most recent Chinese ballistic missile and is considered a full-range ICBM. Development, which may have been delayed by the cultural revolution, began about 1973. The CSS-4 was tested in 1980 and may be just entering operational service. The CSS-4 is a three-stage solid-propellant ICBM with inertial guidance and an accuracy of 0.50-NM CEP. It has a warhead of four MT, and its range may be 6500 NMS. Like the CSS-3, it is silo based.

The CSS-N-X is the first Chinese sea-launched ballistic missile. Development began about 1975 and it was tested in 1982. The CSS-N-X is a solid-fuel two-stage missile with inertial guidance. It is similar to the first Polaris SLBM and may have a range of 1600 NM. Yield could be in the 1.0-MT range. Accuracy could be 0.50-NM CEP. The CSS-N-X (tested on a G class SSB) is likely to be based on new types of Chinese nuclear submarines that could carry 6-12 SLBMs. Two may already be available with three more under construction.

4. SOVIET STRATEGIC FORCES

Soviet firepower ready for the parade through Red Square on 7 November 1963, with rocket artillery (foreground), Frog missiles (center), and SA-1, SA-2 and SA-5 SAMs (right).

Long-term missile programs have been the main factor in development of the Soviet strategic force posture since the end of World War II. New weapons and modifications of old ones have continually been introduced, as military requirements have changed. During the 1950s regional strategic requirements were clearly dominant in decisions on the development of Soviet weapons. But in the early 1960s the intercontinental requirements of Soviet strategic forces increased so dramatically that they came to match regional requirements in importance and urgency. In its efforts to meet the needs defined during the 1960s, the Soviet Union alternately deployed new generations of intercontinental and regional nuclear missiles and sought to acquire variable-range missiles that could satisfy the requirements of both forces.

The Initial Postwar Period

Despite Stalin's public expressions implying a lack of appreciation for the importance of modern weapons, the USSR was vigorously pursuing a variety of short- and long-term programs to improve its military position. By 1949 the Soviet Union had tested its first atomic device, and by 1953 a thermonuclear device had been detonated. Experimental testing of short-range land- and sea-launched missiles was undertaken at the end of the war, and by 1950 design plans for developing medium- and intercontinental-range ballistic missiles (MRBMs and ICBMs) had been established.

Among the first Soviet nuclear weapon systems was a series of tactical missiles – the SS-1 and SS-2 – developed for the Ground Forces in the early 1950s. By 1955 an operational-tactical missile, the SS-3 Shyster,

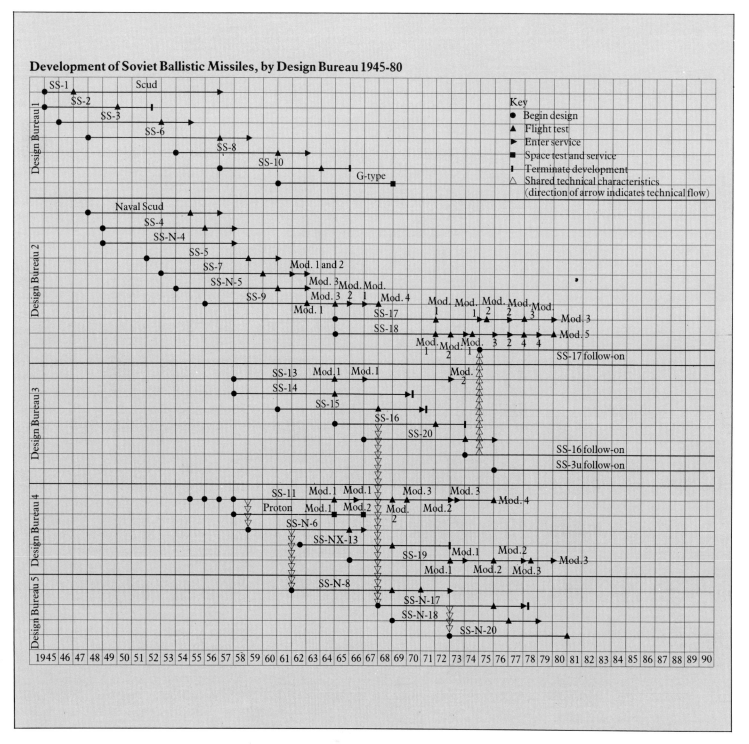

Development of Soviet Ballistic Missiles, by Design Bureau 1945-80

Above: An SS-5 Skean trundles through Red Square on May Day 1968.

was assigned to support Ground Force operations. The missiles' limited range and their command by the Ground Forces suggest that these nuclear missiles were seen as modern equivalents of traditional Soviet artillery in providing fire support. In wartime such missiles would probably have been used to provide long-range barrages of nuclear or chemical explosives to support Ground Forces assaults. The importance of operational-tactical missiles grew through the 1950s as NATO's deployments of nuclear-capable artillery and tactical nuclear missiles expanded.

At about the time that the Soviet Union began assimilating nuclear systems into its force structure, important changes took place in the West's strategic forces that dramatically altered the USSR's regional requirements.

Establishing the Strategic Rocket Forces
By the late 1950s Soviet military doctrine had begun to shift, recognizing that the advent of long-range, nuclear-armed missiles had created a 'revolution in military affairs.' The USSR conducted the world's first full-range test of an intercontinental ballistic missile – the SS-6 – in August 1957, followed two months later by the dramatic launch of the Sputnik satellite into orbit using

a similar SS-6 booster. About the same time, the USSR's first regional-range strategic missile, the SS-4, entered into flight testing. As these strategic missiles neared deployment in 1958-59, difficult questions undoubtedly arose concerning their wartime role and who would command them. These momentous events foreshadowed important changes in the Soviet Union's military force structure and international political position. With strategic missiles, the USSR acquired a technology that could support its regional political and military objectives and directly offset American intercontinental strategic might.

Organizationally, the major result of the shift in doctrine was the establishment in December 1959 of the Strategic Rocket Forces, with responsibility for both the regional and intercontinental missile forces. The SRF was declared to be the preeminent service in wartime, displacing the Ground Forces. Similarly, Soviet industrial resources shifted from strategic-bomber to missile production. Development of the heavy bomber ended with the Mya-50 Bounder, which was first flown in 1957 and never produced; the Bounder may have been developed only as a hedge against the failure of the ICBM program.

Although the change in the Soviet strategic posture was quite apparent, the motivations and methods involved were not. Those who sought both an independent and a central role for the strategic missile forces undoubtedly held complex and even conflicting aims.

The serious strategic challenge posed by Western regional nuclear forces and the prospect of growing US intercontinental forces could probably best be countered by nuclear missiles. The ballistic missile's ability to reach enemy targets over almost any distance in a relatively short time made it far more appealing than strategic bombers, with their lengthy flight times and vulnerability to air defenses. Many Soviet military leaders may have viewed missiles as a modern means of performing the artillery's traditional counter-battery role through decisive strikes against the enemy's main forces. Land-based missiles probably also appealed to both political and military leaders because of their more economical operating requirements and the possibility they afforded of tight command and control.

The Soviet leadership was disposed to play down the importance of the strategic bomber in part as a way of undercutting the significance of the substantial American advantage in intercontinental bomber forces. And Khrushchev, to enhance the Soviet strategic deterrent and to extract concessions in political negotiations over such critical issues as Berlin, was able to sustain the image of a growing Soviet lead in missile technology through frequent misleading claims of Soviet missile strength. The continuing credibility of Soviet claims of technological superiority, however, rested on a series of spectacular achievements in the Soviet space program, not on the less visible ICBM program.

For several years after the Strategic Rocket Forces were established, they were armed predominantly with regional-range strategic missiles. In part, this resulted from the technical difficulties associated with the early Soviet ICBM program. But it also reflected the priority the USSR continued to accord to political and military objectives in the regional theaters surrounding it. The strategic missiles deployed during the late 1950s and the early 1960s were effective means of meeting Soviet regional military requirements that had evolved through the 1950s. The SS-4, a medium-range ballistic missile with about twice the range of the Ground Forces' SS-3 missile, was first deployed in late 1958. By the mid-1960s nearly 600 SS-4 launchers were deployed. The SS-4s were eventually supplemented by about 100 SS-5 intermediate-range ballistic missile (IRBM) launchers, which first became operational in 1961. Together, these missiles could attack a wide range of military and industrial targets throughout Europe and over much of the Far East. Priority targets probably included Western air bases, Thor and Jupiter missile sites and nuclear storage sites. At the same time, the Soviets probably underestimated the pace of the US ICBM program, which got off to a slow start. Using concurrent development programs, the United States rapidly advanced to the deployment of advanced ICBMs, including the solid-fuel Minuteman missile.

By comparison with US missiles, the first generation of Soviet intercontinental missiles did not provide an adequate military capability for the USSR. Despite its worldwide notoriety, the SS-6 Sapwood ICBM was never deployed in more than token numbers. As a military system, it suffered from a series of shortcomings. But as the main booster for the early Soviet space

program, it performed quite well. Hundreds of the SS-6 or its derivatives were produced for the space program, in part because it was probably designed by S. P. Korolev, a pioneer of modern Soviet rocketry who was most closely associated with that program. The space achievements through the early 1960s were clearly valuable to Khrushchev in his political campaign against the West.

By the late 1950s a number of new strategic missiles were in process of development – the SS-7, SS-8 and SS-9, all large ICBMs suitable for covering such area targets as bomber bases and early missile sites on the US mainland. Design plans for other missiles were probably being set. These included the smaller SS-11; the USSR's first solid fuel ICBM, the SS-13, and the *Yankee*-class submarine and its ballistic missiles.

Changing Requirements and Expedient Solutions
Development of the Soviet intercontinental missile force began in earnest in the 1960s. Changes in the military requirements of Soviet intercontinental and regional forces, as well as technical shortfalls in key missile programs, compelled the USSR to redirect many of its programs and devise expedient measures. Important among these was the reorientation of the SS-11 ballistic missile from an anti-naval role to that of a system flexible enough to perform both regional and intercontinental strategic missions.

The rapid expansion and alterations in the United States' plans for its intercontinental strategic forces that occurred during 1961 radically affected the US-Soviet

Left: Launch of an SS-4 Sandal from a surface pad.
Below: The SS-8 Sasin ICBM was first seen in a Red Square parade in 1968.

military balance. Having entered office claiming the existence of a missile gap favoring the USSR and the need for a greater range of military options, the Kennedy administration promptly accelerated and redirected several major defense programs in its first year. Authorizations for the Polaris program leapt from 19 to 41 submarines, with delivery schedules significantly advanced; the number of Minuteman ICBMs planned for deployment in hardened silos nearly tripled and plans for a rail-mobile system were canceled. Emphasis on strategic bombers disappeared with the cancellation of the B-70 high-altitude-bomber program. In less than a year's time, it became evident to the USSR that it was being rapidly outpaced by the United States – both qualitatively and quantitatively – in terms of intercontinental strike forces.

The USSR suddenly faced a serious threat from American forces that could not be directly countered by either its growing regional missile force or its strategic defensive forces. The initial burden of responding to the American ICBM force would fall on the second-generation Soviet ICBMs, the SS-7 Saddler and SS-8 Sasin, which would not be available for deployment until 1962-63. Their relatively poor accuracy and high vulnerability to counterattack made them an inadequate counter, however, to the new Minuteman generation of missiles. Furthermore, the low production rate of these large Soviet ICBMs would permit the United States to open up an expanding lead in number of missiles deployed. Beyond that, Soviet anti-submarine and ballistic-missile defenses were of limited utility. By 1961 the Soviet political and military leadership undoubtedly realized that its planned strategic posture was inadequate to meet the evolving American strategic threat.

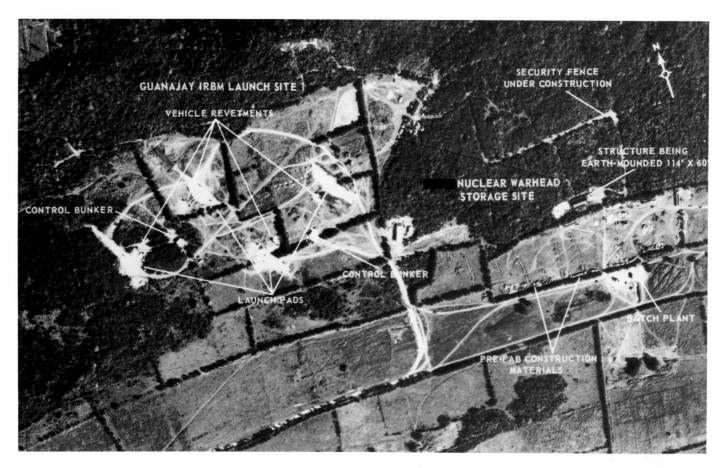

The pressure on the Soviet leadership to eliminate the American strategic advantge, and the years required before a new generation of strategic missiles could be deployed, apparently led Khrushchev to gamble on placing Soviet regional-range missiles on the island of Cuba. By September 1962 the first strategic missiles had been sent to Cuba in cargo ships. The Soviet plan apparently was to deploy some 24 SS-4 medium-range and 16 SS-5 intermediate-range missile launchers (each with one reload) to ten separate sites. Consistent with Soviet all-arms thinking, the missiles would be only part of a package including air defenses, Soviet ground forces and local security forces. Soviet MRBMs and IRBMs based in Cuba could cover a variety of important US strategic targets like ICBM and bomber bases as well as command-and-control sites, although the soft basing sites of the Soviet missiles were vulnerable to attack in wartime.

This forward deployment of its strategic missiles probably offered the USSR its only means of quickly improving its strategic position relative to the United States. But it was a very high-risk venture, and it ended in a major confrontation with the United States. During the course of the October 1962 crisis, the United States backed its demands for the withdrawal of the missiles with a naval quarantine of Cuba and compelled the USSR to dismantle and withdraw its nuclear missile launchers. The failure of Khrushchev's gambit in Cuba meant that the Soviet Union's response continued to rest on its long-term intercontinental force program.

Most of the Soviet ICBMs deployed or under development in 1961 were designed to destroy such area targets as bomber bases, early US ICBM sites or administrative and economic centers. The US Minuteman,

then under development, was not only to be deployed in hardened underground silos, but it had a relatively high degree of accuracy which made it a formidable threat to the early Soviet missile forces.

Given the years it would require to develop new strategic weapons more appropriate to countering the US forces, the USSR had to rely on its existing missile programs. Shortcomings in its ballistic-missile defenses placed the entire burden on its offensive forces. The third generation of Soviet ICBMs, to be deployed in the mid-1960s, offered the only solution. It included three new ICBMs (the SS-9, the SS-10 and the SS-13), each produced by a different design bureau. The USSR apparently turned to the SS-9 and the SS-13 ICBMs (the SS-10 was later canceled).

The SS-9 Scarp, produced by the Yangel design bureau to replace the SS-7 and to cover large-area targets, possessed greater accuracy and a larger yield than the SS-7, making it well suited for attacking hardened point targets such as Minuteman missile silos or underground launch-control centers. But the size and

Left above: US reconnaissance photograph of a Cuban missile site, said to be for SS-4 MRBMs, near Guanajay, on 17 October 1962.
Left below: This Vostok space launcher, with 32 first-stage rocket engines, was based directly on the SS-6 Sapwood, the world's first ICBM.
Above: An SS-10 Scrag ICBM seen in Red Square in November 1968.

expense of the SS-9 probably precluded production at levels needed to match the high deployment rate of the Minuteman force. Consequently, the USSR was compelled to diversify its approach to the problem. It apparently planned to deploy enough SS-9s to neutralize the US Minuteman force through attacks on the one hundred very hard launch-control centers critical to its operation, while counting on the deployment of substantial numbers of SS-13 ICBMs to match politically (and perhaps militarily counter) the Minuteman missile for missile.

The SS-13 Savage program represented a significant departure from the earlier Soviet ICBM series designed for attacking large-area targets. Unlike its predecessors, the SS-13 utilized solid-fuel propellants and carried a small nuclear warhead. In many repects this missile, developed by the Nadiradize design bureau which specialized in solid-fuel missiles, appears to have been intended as the Soviet counterpart to the American Minuteman I; it was similar in size and propellant type and started development about the same time as the Minuteman in the late 1950s. The smaller size of the SS-13 could have allowed it to be produced more rapidly and cheaply than the larger Soviet missiles. The SS-13 seems to have been intended to serve also as a strategic reserve for the Soviet land-based missile force. Its propulsion system made it feasible to deploy it as a mobile missile as well as difficult to locate and destroy it in wartime.

Serious technical problems associated with the SS-13's guidance system and solid-fuel motor apparently hampered its development. Only sixty of the ICBMs were deployed, all of them in silos. One of the important repercussions of this failure may have been Soviet rethinking of the nature of the future strategic reserve force. The immediate response, however, was the search for a replacement.

The Soviet leadership found its substitute by turning to a new source, the missile design bureau of V. N. Chelomei. One of its products was the SS-11 Sego, which would eventually satisfy the Soviet need for an ICBM that could be deployed promptly enough to match the US Minuteman missiles in number. Like the SS-13, the SS-11 was relatively small, with modest accuracy and a small warhead, but it used storable liquid fuel. The SS-11 seems also to have been a program that was redirected to meet the new military requirements. And it had a variable-range capability, which enabled it to strike targets at less than its full intercontinental range.

The many anomalies associated with the program suggest that relatively late in its development cycle the SS-11 was selected as a means for matching US Minuteman deployments. Given the naval heritage of its design bureau, it is possible that the SS-11 was originally designed in the mid-1950s as a land-based ballistic missile for use against enemy naval forces at long distances. Statements and technical developments at the time provide evidence of Soviet interest in developing a weapon to counter the significant threat posed by nuclear-armed, carrier-based aircraft in the American strategic strike force. The difficulty of redirecting the development of the SS-11 was probably eased by the fact that the US program that created the need for the missiles also downgraded the role of the aircraft carrier as American ballistic-missile submarines became operational. The combined failure of the SS-13 program and the rapid change in the nature of the strategic threat in the early 1960s were only the first in a series of new requirements that affected the development of the SS-11 missile.

Evolving Roles for the Sea-based Forces

Important decisions were also made during the early 1960s about the strategic missions of the Soviet Navy. With the formation of the Strategic Rocket Forces, the Navy apparently lost any major role in conducting strategic strikes (less than one hundred sea-based ballistic missiles were operational through the mid-1960s).

In the late 1960s the USSR also began expanding its own sea-based ballistic-missile forces. In 1968 Yankee Class submarines were armed with the SS-N-6 Sawfly missile. While addition of the SS-N-6 generally enhanced the Soviet Union's overall strategic position, the Yankee Class submarine was apparently assigned only a limited role in intercontinental strike missions against targets located in US coastal areas, probably in part because its survivability in wartime is questionable. Nonetheless, deployment of these submarines signaled the beginning of a period in which the role of Soviet sea-based ballistic-missile forces would increase, especially in the intercontinental theaters of operations. Construction of strategic-missile submarines had the highest priority in the Soviet naval program, thus signifying the Soviet leaders' sense of urgency in enhancing the sea-based missile force and matching the US sea-based forces.

Despite the growing importance of the nuclear-armed submarine within the Soviet force structure and the belated recognition of its usefulness in strategic strikes, there is reason to believe that the original requirement for the Yankee Class nuclear-powered ballistic-missile submarine (SSBN) was to defend against enemy naval forces. A combination of changing national requirements and technical difficulties altered its role as it approached deployment. When this submarine entered development in the late 1950s, the Soviet Navy's primary mission was to defend against Western sea-based nuclear forces. With the creation of the SRF, the Navy's role in the USSR's strategic-strike plans was tentative at best. Shortly after work began on the Yankee SSBN, a tactical ballistic missile – the SS-NX-13 – apparently began development. This missile could have been launched from the Yankee. The SS-NX-13, with a 400-mile range and a terminal guidance system, would have been suitable for use against Western carrier task forces, and possibly even against Polaris submarines.

Soviet Navy plans to deploy the Yankee SSBN in this role appear to have been altered by the technical difficulties that plagued development of the SS-NX-13 and precluded its deployment. But the Yankee program was most affected by the changes in intercontinental requirements that Soviet leaders began to confront in 1961. They were apparently convinced of the need to expand the Soviet sea-based strategic capability and politically match the growing US SSBN force. Sea-based missiles not only would improve the survivability of the strategic forces, but could also serve as a strategic reserve.

Initial evidence of this shift in Soviet thinking was the regular assignment by 1964 of older Soviet ballistic-missile submarines to patrol in the open ocean. Within a few years the first Yankee Class SSBNs were deployed, carrying the SS-N-6, a missile with a range of 1300 NM

designed by the Chelomei bureau from existing components. By 1970 patrol areas for the Yankee submarines were established off both US coasts, thereby giving the USSR sea-based coverage of a wide range of targets. Nevertheless, during this period the Soviet SSBNs were apparently limited to covering US SSBN support and communication facilities, large ports and major US home fleet centers like Norfolk, Virginia. Soviet writings indicate that nuclear strikes against such targets would be considered an extension of the Soviet Navy's wartime missions of combating the enemy's fleet and disrupting his sealanes of communication.

The ability of Yankee Class SSBNs to serve as a strategic reserve was probably questionable. Unlike American submarines, which were considered highly survivable when at sea, the Yankee SSBN had a doubtful chance of surviving long in a US-Soviet conflict. The Soviet submarines had to slip through Western-controlled choke points to reach their patrol areas, and because they were relatively noisy were vulnerable to

Top: An SS-N-6 Sawfly submarine missile on parade through Red Square.
Above: A Soviet Yankee class missile submarine photographed in early 1976.

detection. In addition, the relatively short range of the SS-N-6 missiles limited the ability of these submarines to perform deep-strike missions against all US strategic targets without risking detection. But the SS-N-6s were probably quite useful for covering regional targets; since only a small part of the Yankee SSBN force was normally on patrol near US coasts, many of these submarines could have been on call in home waters for regional support.

Problems for the Regional Nuclear Force

The shift of Soviet resources in the mid-1960s to building up the intercontinental force did not signal a decline in the importance of the regional-range forces. Instead, it reflected the immediacy of the need to offset the

expanding American intercontinental threat. In the mid- to late 1960s, new threats to the survivability of the Soviet regional forces prompted the USSR to turn to variable-range weapons to ensure the effectiveness of its regional strategic posture. By the mid-1960s, about 700 SS-4 and SS-5 ballistic missiles had been deployed in the Soviet Union, most of them in clusters at soft sites. When the silo-hardening program ended in 1964 with the shift in emphasis to the ICBM force, much of the Soviet regional nuclear forces was left highly vulnerable. Previously, the primary threat to these forces had been from strategic aircraft or nuclear cruise missiles, which were relatively slow and could be intercepted by Soviet air defense, or possibly even preempted by a nuclear missile strike. This situation changed dramatically when the United States deployed long-range ballistic missiles – the quick-reacting and highly survivable Polaris missiles in particular could effectively strike both Soviet regional nuclear missiles and strategic bomber bases. As early as May 1963 the United States had committed five Polaris submarines to NATO's defense, and in 1964 the Polaris A-3, with its greater accuracy and multiple warheads, became operational. The long-range Mace surface-to-surface cruise missile deployed in Germany could also reach Soviet regional forces. Beyond that, in the early 1960s the Soviet Union faced the prospect that the United States and its NATO allies might deploy a land-mobile, medium-range ballistic missile.

The USSR undoubtedly planned to protect its regional nuclear forces by deploying mobile ballistic missiles. By the mid-1960s testing began on two such missiles, the medium-range SS-14 and the intermediate-range SS-15. Soviet defense planners probably envisaged them as a more survivable replacement or supplement to the increasingly vulnerable SS-4s and SS-5s. However, technical problems arose with the development of these expensive, solid-fuel missiles. To compound these problems, a new set of targets for the Soviet regional forces emerged in the form of China. Sino-Soviet relations had fully deteriorated by the mid-1960s, and the Soviet military – presumably for the first time – received new requirements to cover Chinese targets. By 1969 additional Soviet nuclear forces were deployed close to the border.

With the survivability of its forces decreasing and its target requirements expanding, the Soviet Union apparently turned again to the SS-11 missile. From a regional-range missile for striking naval targets, it had been redirected to help meet intercontinental requirements. Just as it neared deployment as a long-range weapon, the SS-11 system was probably drawn upon once more to serve as a more survivable system for Soviet regional forces. With its variable range, and deployed in hardened silos, the SS-11 was well suited to its expanded role. The first SS-11s became operational in 1966; by 1968 deployment had begun, with 120 missile launchers at sites in western Russia previously used only for regional missiles, while other SS-11s deployed near the Sino-Soviet border were probably assigned to cover targets in the Far East. The significant contribution of this new component of the regional strike forces meant that the USSR gave up little in 1970-71 when it dismantled about 70 of the more vulnerable SS-4 and SS-5 launchers in the Far East. In combination with the submarine-launched SS-N-6, the SS-11 would have made a critical contribution to the overall survivability and target coverage of the Soviet regional forces when they were under great stress.

The Contemporary Period

With the deployment of the fourth generation of regional and intercontinental missiles in the 1970s, substantial improvements were made in the Soviet strategic force posture. The USSR's third-generation missile deployments had matched the US missile force in number, and in the following generation significant advances were made in the overall effectiveness and survivability of the force. Paradoxically, the USSR had attained significant success in fulfilling its political aims in the buildup of its strategic force in the 1960s but had generally failed to meet its military requirements. It was this unusual combination of circumstances that set the stage for the

initiation of serious US-Soviet strategic arms limitation talks (SALT I) which resulted in 1972 in the signing of the Anti-ballistic Missile (ABM) Treaty and an interim agreement limiting strategic offensive arms.

Creating a Strategic Reserve Force at Sea

At some point in the mid- to late 1960s the Soviets apparently decided to diversify their intercontinental strategic strike forces by the development of long-range sea-based missiles. This national decision to augment the silo-based ICBM force with a substantial strategic reserve at sea eventually required major alterations in the role and character of the Soviet Navy. Coupled with the deployment of Delta Class SSBNs armed with long-range sea-launched ballistic missiles (SLBMs), such a decision meant that the Soviet submarine force was finally to share in the deep-strike (or strategic intercontinental strike) mission against a wide range of targets in the United States. To ensure the survivability of its SSBN force, the Soviet Navy added a new mission, the strategic support of the SSBN force. Surface ships and

other forces would be used to defend the force in wartime against Western threats.

Beyond the inherent advantages of a sea-based ballistic missile, the failure of the SS-13 program to furnish a land-mobile ICBM during the critical years of the mid-1960s probably caused the Soviet leadership to decide to enlarge the sea-based missile force. The timing of the SS-N-8 SLBM's development and the nature of its progressively lengthened flight-test series suggest that it was something of a 'quick fix' measure. So does the fact that the missile is launched from a Delta Class SSBN, which is simply a modification of a Yankee-class hull rather than a new design. (In 1978 a much-improved missile with multiple warheads – the SS-N-18 – was deployed on the Delta III Class SSBN.)

The deployment of submarines with longer-range missiles beginning in the early 1970s reflected an evolution in the Navy's mission as well as the Soviet Union's decision to move a large part of its strategic forces to sea. The Soviets increasingly attributed to their SLBM force the capability to destroy land targets both on the enemy's seaboard and deep in his rear. Until the early 1970s the SLBM force was probably not recognized as being prepared to perform the full strategic strike role against the enemy – a mission to which the American SSBN force has been dedicated from its inception.

Major Advances with Fourth-Generation Deployments

With the deployment of its fourth-generation ICBMs during the 1970s, the Soviet Union made impressive gains in the overall capability of its land-based missile force. These ICBMs, designed in the mid-1960s to replace existing missiles, were expected to meet the SRF's diverse targeting requirements and to enhance the forces survivability. The new deployments offered an increasingly effective (and redundant) capability for countering the US ICBM force. The large SS-18, put into operation in late 1974 to replace the SS-9, was

Below: The Delta III class submarines are each armed with 16 of the formidable SS-N-8 missiles.
Left: A Delta I submarine, showing the high missile bay for 5,700-mile SS-N-8 SLBMs

Average Deployment Rates of Soviet Intercontinental Forces

Type	Years Deployed	Average Annual Deployment Rate
SS-7/8[1]	1962-65	55
SS-9	1966-70	58
SS-11[2]	1966-71	142
SS-13	1967-71	12
SS-17	1975-80	25
SS-18	1974-80	61
SS-19[3]	1974-81	23

[1] Not including SS-7 and SS-8 reloads. If these missiles were also included, the SS-7/8 deployment rate would be 91 per year.
[2] Not including SS-11s deployed in regional missile fields. If these missiles were also included, the SS-11 deployment rate would be 162 per year.
[3] Not including SS-19s deployed in regional missile fields. If these missiles were also included, the SS-19 deployment rate would be 46 per year.

capable of attacking very hard targets. One version of the SS-18 with a single large warhead was suitable for attacking such targets as US Minuteman launch-control centers, while other SS-18 ICBMs carrying multiple warheads made it possible for the USSR effectively to extend its target coverage to the Minuteman silos for the first time.

The Minuteman force was not the SRF's only targeting responsibility against the United States. The SS-17, with four independently targeted, large-yield warheads, probably supplanted the SS-7 in coverage of air bases and other area targets. Both the new SS-19 ICBM and the third modification of the SS-11, which carried multiple warheads, were capable of covering various targets in regional and intercontinental theaters. The SS-19 was also accurate enough to threaten hardened targets and thus provided a backup to the SS-18 force. A variety of modifications to these systems, such as single-warhead missiles, improved the Soviet capability against special kinds of targets in very hard silos or those that were not easy to locate.

In terms of improving ICBM survivability, the USSR undertook a vigorous program to harden its silos, probably in the expectation that the United States would deploy a new ICBM by the late 1970s. Public reports suggested the United States might develop a more accurate missile with a large throw weight, known as the WS-120A, which would become operational soon after deployment of the Soviet fourth-generation missiles. The continuing threat to the Soviet land-based missile force probably helps to account for initiation of the SS-X-16 mobile missile program in the fourth generation of Soviet ICBM development. The SS-X-16 was lighter than its contemporaries and like its predecessor, the solid-fuel SS-13, could deliver only a relatively small payload. By the early 1970s it became clear that the United States would not deploy a new ICBM for another ten years, and that the Soviet silo-based ICBM force was sufficiently survivable to preclude the need for a mobile ICBM. Given the SS-X-16's limited capability and reported testing difficulties, the USSR had little incentive to deploy more than a small force at the expense of its silo-based ICBMs. Consequently, it agreed to prohibition of the SS-X-16 in the SALT II negotiations because of its similarity to the intermediate-range SS-20. It may have seemed preferable to the USSR to delay deployment of a solid-fuel mobile ICBM until a more effective model could be developed.

By the late 1970s the USSR had also modernized many of the initial weapon systems assigned to its regional nuclear forces. One of the first new systems was the Backfire bomber (replacing the aging Badger bomber) that became operational in 1974. The Backfire is effective against both land and sea targets because it can fly at low altitudes as well as high speeds and thus reduce the amount of lead time for enemy air-defense systems to react. It has the flexibility to perform both nuclear and non-nuclear missions. Deployment of the SS-19 and SS-20 land-based ballistic missiles added substantially to the capability of regional nuclear forces. The SS-19 provided a relatively secure land-based reserve force for covering regional as well as intercontinental targets. The SS-20 IRBM – a long overdue follow-on to the vulnerable SS-4 and SS-5 missiles – with its multiple warheads greatly increased the regional forces' target effectiveness and coverage. Its solid-fuel propulsion system enables the SS-20 to be deployed as a mobile missile, which improves its survivability and makes this missile a major addition to the Soviet regional nuclear capability.

The regional forces' flexibility has been significantly improved with deployment of a series of new shorter-range missiles (the SS-21, SS-22 and SS-23). Increasingly, other new systems are capable of performing missions similar to those of the Soviet regional strike forces. The Su-24 Fencer, for example, is an attack aircraft that can adequately perform non-nuclear and nuclear attacks which could previously be undertaken only by medium bombers. With technological improvements of this kind, the earlier distinction of roles between Soviet tactical and regional forces has steadily diminished.

Prospects for the Next Generation
The fifth-generation Soviet missile deployments seem likely to feature improvements in accuracy and attempts at dealing with long-term threats to missile survivability. The new designs and modifications to current ICBMs nearing the flight-testing stage could include both liquid-fuel missiles and solid-fuel missiles. The new missiles are probably intended to replace the older silo-based SS-13 and SS-11 ICBMs. The Soviets may also develop new mobile SS-X-16 ICBMs.

The new ICBMs are likely to have the accuracy necessary to a high probability of destroying a hardened enemy target with only a single warhead. While it involves some risk of technical failure or delay, Soviet transition to solid-fuel missiles could offer greater missile efficiency, safer handling and lower operating costs. The fact that large numbers of liquid-fuel missiles will remain in the SRF inventory for some time helps minimize the risks associated with attempts to deploy solid-fuel missiles, an area that has proved difficult for the USSR.

During the 1980s the USSR also must be concerned with the growing capability of US strategic missiles (Peacekeeper ICBM, Trident II SLBM and the long-range cruise missile) to threaten the survivability of its land-based ICBM force. With one exception, few major changes are likely to occur in the Soviet Union's ICBM deployment patterns. The USSR has apparently been taking measures to mitigate the vulnerability of its ICBMs. As part of that continuing effort, it will have to harden its new launchers if they are to be adequately protected.

The deployment of long-range SLBMs on Delta Class ballistic-missile submarines is a major contribution to the survivability of the Soviet missile force. This helps to explain the Soviet willingness to agree in SALT I to dismantle the older SS-7 and SS-8 ICBM launchers of the Strategic Rocket Forces in exchange for additional SLBM deployments. Since the Soviet Union has already made the decision to move to the sea, it is unlikely that it will deploy many more SLBMs even if the vulnerability of its silo-based ICBMs should increase in the future.

Improvements in Soviet early-warning systems and missile-alert rates since the late 1960s raise the possibility that launch under attack has become a basic element of Soviet operational doctrine in the event of a US-Soviet nuclear conflict. Nevertheless, the silo-hardening program continues, suggesting that the Soviet leadership is not content to rely simply on this operational option.

The one new development that is likely for improving Soviet ICBM survivability is the deployment of mobile ICBMs, as a supplement to a larger silo-based force. The US cancellation of plans for an improved replacement for the Minuteman III ICBM (in addition to possible technical problems in the SS-X-16 program) probably enabled the USSR to forego deployment of mobile ICBMs, in accordance with the SALT II agree-

Left: SS-1 Scud missiles elevated to the launch position on their IS-3 tank chassis.

ment. The requirement for deploying mobile ICBMs in the next generation of missiles, however, will be more compelling. They can serve as a hedge against an unexpected vulnerability in other components of the strategic missile forces. While mobile ICBMs are expensive to maintain, Soviet deployments are likely to avoid complicated basing schemes and to be limited in number because of Soviet reluctance to give up the close control of the silo-based force.

The USSR is also likely to upgrade its strategic reserves in the early 1980s. Currently it is testing a long-range solid-fuel SLBM, the SS-NX-20, which will probably be more effective than the SS-N-8 and the SS-N-18 sea-based missiles. Deployed on the large Typhoon Class SSBN, the SS-NX-20 is a major step toward a highly survivable strategic reserve, which the USSR has been seeking since the late 1960s.

Soviet Battlefield Missiles

The work on tactical-range missiles in the post-war period provided the modern foundation for a number of such missiles that would enter Soviet service over a period of years. These included the Scud A, deployed in 1957; the SS-1C, deployed in 1965, and the follow-on SS-X-23; the FROG missile family from 1956 to 1967, and its replacement, the SS-21, as well as a wide variety of other naval and air-delivered missiles. This type of development is apparently an on-going process that responds primarily to changes in Soviet tactical doctrine and in the technological 'state of the art.'

Equally important was the post-war development work on such early missiles as the SS-1 Scunner and the SS-2 Sibling. These missiles were derived from German efforts at the end of the war, and their designs provided the operational 'know-how' for much longer-range missiles still in the concept and development stages. They were both based on German V-2 technology. Although tested in 1947 and 1950, it is not believed that these missiles were in actual service in the early years; rather they served as research and development test beds and crew-training platforms. The SS-1 and SS-2 used liquid-fuel propellant and autopilot guidance. The SS-1 evolved through three versions, the final one having two and one-half times more thrust than the original V-2. The SS-2 retained this increase in thrust, but because of extra fuel tanks had longer range than the SS-1.

Operational-Tactical Missiles

Operational-Tactical Missile (OTM) range falls between those offered by tactical and strategic missiles. As a result, OTMs offer a certain flexibility to Soviet tactical planners. These missiles, now all mobile, are assigned to the higher command echelons such as fronts, unlike Soviet tactical missiles. As a result, they are removed from direct contact with the enemy and more likely to survive the initial phases of a conflict as compared to nuclear-capable aircraft. While possessing potentially greater survivability, their response time and nuclear warhead size (or special non-nuclear munitions) make them well suited to resolve a stalemated tactical operation hundreds of miles away in favor of Soviet divisions or naval formations.

Operational tactical missiles have received only modest attention from Soviet designers. This may result from the effectiveness of other weapon systems, such as cruise missiles for the anti-carrier role, or from the fact that the systems now deployed are adequate to determine the outcome of a tactical engagement whenever it is required.

The SS-3 Shyster was a first-generation operational-tactical missile operated by the Soviet Ground Forces. As a follow-on to the SS-1 and SS-2, this missile represented the next logical step toward longer-range missile systems. Testing began by 1954 and deployment started in 1955. The missile was liquid-fueled and relied on radio-command guidance. With a range of 600 NM, it could carry either a chemical or 200-kiloton nuclear warhead. It is possible that up to 100 launchers were in service through the 1960s. The SS-3 was a transportable weapon launched from a simple gantry and pad complex.

The SS-12 Scaleboard was a third-generation OTM operated by the Soviet Ground Forces as a replacement for the SS-3. It first became operational in 1965 in the western USSR and in 1970 in the Far East. The missile has an inertial guidance system and a range of about 600 NM. Its warhead yield is about 500 kilotons. At one time 72 launchers were deployed on trucks. The SS-12's mobility eliminated the major drawback of the SS-3's vulnerability in a combat environment. The SS-12 would be employed by an army front commander some 200 miles away from the forward battle lines to deal with unfavorable ground combat situations, independent of the Strategic Rocket Forces' strike plan. It is expected to be replaced by the new SS-22 missile, which is also mobile.

With certain caveats, the framework used here to categorize land-based missiles can be also applied to

Above: Frog 7 rocket on its ZIL-135 cross-country launch vehicle.
Below: An SS-12 Scaleboard on its large MAZ-543 transporter/erector/launcher, photographed in 1978.

Left above: Soviet artillerymen use a gyrocompass to establish the launch azimuth (bearing) for an SS-1 Scud missile.

sea-based missiles. One limitation on so doing arises from the fact that the mobility of sea-based launchers enables these systems potentially to perform variety of wartime roles. For instance, short-range SLBMs like the 350-NM SS-N-4 could be forward deployed to cover such intercontinental targets as US naval bases. Therefore, the submarine launcher's mobility prohibits a strict definition of the SLBM's role based solely on its range capability.

Initial development of a submarine-launched ballistic missile capability by the USSR began in the early post-war years as the Soviets explored the feasibility of previous German experiments in this area. The aborted German program for V-2 ballistic missiles to be launched from sea-going containers, towed underwater by a submarine and uprighted for launching, is one example. Tests of this system in the early 1950s with advanced Soviet versions of the V-2 known as the Golem missile series proved this launching system to be unfeasible and led the Soviet Union to develop a missile that would be launched from the submarine itself. In September 1955 the USSR achieved the first launch of a ballistic missile from a submarine. The missile utilized in the early Soviet test was actually a short-range ground force weapon, the Scud, which was launched while the submarine was on the surface.

Following the naval Scud, the USSR developed its first-generation sea-launched ballistic missile, the SS-N-4 Sark. This two-stage liquid-fueled missile was a major improvement with a range of 350 NM and a warhead yield of 2 to 3.5 megatons. Initially tested in 1957, the SS-N-4 was first deployed on the Golf Class SSB in 1958 and the Hotel Class SSBN in 1959.

The other Soviet SLBM that falls within the operational-tactical missile range category is the SS-NX-13. This third-generation SLBM was probably never operationally deployed. Series testing of this missile appears to have extended from the late 1960s and terminated by 1974. Guided to its target by a semi-active terminal homing device, the SS-NX-13 had up to a 400-NM range and the payload capacity for a 700-kiloton warhead. Given the SS-NX-13's similarity in size to the SS-N-6 SLBM it is possible that it was intended for deployment on the Yankee Class SSBN. As a naval operational-tactical missile, it might have been developed to counter aircraft carriers, or even SSBNs.

Strategic Missiles

These missiles are longer-range weapons capable of striking enemy targets of strategic importance to the outcome of a conflict. Such targets are usually found deep in the enemy's rear areas, whether located in the various theaters of military operations (TVDs) adjoining the USSR, or in the broader intercontinental areas. Strategic missiles are subdivided by their range capability into medium-range, intercontinental and global missiles.

Medium-range Missiles

The Soviet Union's medium-range missile development has reflected the evolution of its military requirements in the regional areas surrounding it. Along with

medium bombers, these missiles are responsible for covering a wide range of targets in Europe, the Middle East and Asia.

In many respects the medium-range missiles are distinctive compared to other Soviet strategic missiles. In terms of their basing mode, the initial medium-range ballistic missiles were extensively deployed above ground in soft bases which enabled them to retain a reload capability. Similarly, the only operational deployments of land-based mobile ballistic missiles to date by the USSR have involved missiles in the medium-range category.

The Soviet medium-range missile force is also notable for its reliance on certain variable-range ballistic missile systems that are capable of striking regional as well as intercontinental targets. Although also deployed as ICBMs, the SS-11 missile (and its follow-on, the SS-19) possesses a capability for variable-range targeting. At least one field of these missiles, comprising 180 silos, appears to be assigned to covering a variety of regional Western targets in wartime. Other SS-11s closer to the Chinese border could be considered targeted at China. (Both these missiles are discussed in the intercontinental section.) The other missiles capable of variable-range missions are the medium-range Soviet SLBMs. Of these, the SS-N-5 is discussed in this section, while the SS-N-6 and SS-NX-17 are treated in the intercontinental section.

The SS-4 Sandal was a first-generation MRBM. Initially, this missile may have been operationally controlled by the Long-Range Aviation group until the

Number of Soviet Regional-range Weapons and Warheads, 1955-80

Instrument	1955	1960	1965	1970	1975	1980
Weapons	1320	1580	1718	2060	2219	1232
Land-based missiles	24	248	733	971	990	1032
SS-3	24	48	28	0	0	0
SS-12	0	0	0	54	72	72
SS-4	0	200	608	508	508	360
SS-5	0	0	97	90	90	40
33-14	0	0	0	29	0	0
SS-20	0	0	0	0	0	180
SS-11[1]	0	0	0	290	320	260
SS-10[1]	0	0	0	0	0	120
Sea-based missiles[2]	0	36	105	365	569	445
SS-N-4, SS-N-5	0	36	105	93	89	57
SS-N-6	0	0	0	272	480	388
Bombers[3]	1296	1296	880	724	660	655
Tu-4	996[2]	296	0	0	0	0
Tu-16	300	1000	775	550	475	445
Tu-22	0	0	105	174	170	140
Tu-22M	0	0	0	0	15	70
Warheads	324	1034	2085	2301	2467	3497
Land-based	24	248[4]	733[5]	969[6]	990[7]	1992[8]
Sea-based missiles	0	36	105	365	569	445
Bombers	300	750	1247	965	908	1060

[1] Includes only those missiles facing Western Europe and China.
[2] Those with less than an intercontinental range: does not include forward-deployed units.
[3] Roughly 75-80 percent are strike aircraft.
[4] Conventional capability only.
[5] Plus a reload capability of about 200 SS-4s.
[6] Plus a reload capability at soft sites of about 514 SS-4s and SS-5s.
[7] Plus a reload capability at soft sites of about 466 SS-4s and SS-5s.
[8] Plus 180 SS-20s and a reload capability at soft sites of about 265 SS-4s and SS-5s.

Above: An SS-13 Savage solid-propellant ICBM pictured in the May Day parade in 1967.

creation of the Strategic Rocket Forces in 1959. The SS-4 was first tested in 1958 and deployed in late 1958. Original versions were liquid fueled and had radio-command guidance. Later models produced in the early to mid-1960s had storable liquid fuel and inertial guidance. The SS-4 has a range of 1020 NM and a warhead yield of 2.0 megatons. Other warhead combinations allowed chemical strikes. By 1965 600 SS-4 launchers were in service with 516 at soft missile sites and 84 in hardened launchers.

The SS-5 Skean was a first-generation IRBM and an outgrowth of the SS-4. It was first deployed in 1961. The SS-5 used storable liquid fuel and an inertial guidance system. Its range was some 2200 NM and it carried a large 4.0-megaton warhead. Chemical warhead versions are also likely to have been produced. By 1965 a maximum of 100 launchers was in service. Its soft-site deployment was similar to that of the SS-4 but totaled only 52 launchers. Another 45 were deployed in hard-site configurations.

The SS-14 Scapegoat and SS-15 Scrooge were third-generation solid-fuel ballistic missiles of regional-ranges that were operated by the SRF. Both were based on the SS-13 Savage ICBM and were developed as mobile missile systems. In testing, the SS-14 was a more reliable system, with only two failures out of 16 tests

from 1965-70. In the same period the SS-15 had three failures in eight test shots. Both the SS-14 Scapegoat and the SS-15 Scrooge were mobile missile systems that were never deployed with regular missile units. After a lengthy development period, both served as crew trainers and may have been briefly deployed during the early 1970s to the Far East before they were withdrawn from active use. The SS-14 used the upper two stages of the SS-13 ICBM, while the SS-15 used the first and third stages.

The USSR deployed its first real mobile missile, the SS-20 IRBM, in 1977. A fourth-generation regional-range missile, it is operated by the SRF. The SS-20 serves as a follow-on to the mobile SS-14 and SS-15 missiles. It was first flight-tested in 1975. The fact that the SS-20 is a product of the Nadiradize design bureau and is simply a derivative of Soviet ICBM development raises the possibility that it is only an interim system to a new-generation IRBM. The SS-20 uses the upper two stages of the solid-fuel SS-X-16 ICBM and has an on-board computer with its guidance system. It carries three MIRV warheads, each with a possible yield of 150 to 600 kilotons. The SS-20 is a mobile missile that can move along road networks to increase survivability but is normally deployed in a garrison; it is probably located in soft garages with pre-surveyed firing positions. Some of the garrisons may be located in the vicinity of former SS-7 and SS-8 ICBM bases.

The SS-N-5 Serb is a second-generation SLBM produced for the Soviet Navy. It was first deployed in 1963 and is carried on Hotel II SSBNs and Golf II submarines. This liquid-fuel two-stage missile has a range of 850 NM and a megaton-yield warhead. Unlike its predecessors, the SS-N-5 can be launched from underwater, thereby minimizing the likelihood that the submarine will be detected prior to the missile's launching. It is currently deployed on a small number of remaining Golf II and Hotel II Class ballistic-missile submarines.

The SS-N-6 Sawfly is a third generation SLBM. Unlike other SLBMs, the SS-N-6 appears to be something of a hybrid, which drew upon land-based SS-11 components. Both the SS-11 ICBM and the SS-N-6 appear to share a number of common technical features, such as the mode of their reentry vehicles. The SS-N-6 first became operational in 1968. The missile has two stages and uses liquid fuel; it has a number of different modifications. The Mod 1 had a range of 1300 NM and carried a single warhead of 700 kilotons. Although the SS-N-6 Mod 1 did not provide full target coverage of the continental United States, this shortcoming was later corrected when, in October 1972, the USSR began flight testing an improved version of the SS-N-6 known as the Mod 2 which became operational in 1973. Its range is 1600 NM and it has a warhead yield of 650 kilotons. This increase in missile range enabled each Yankee class SSBN to cover any target in the US from the 100-fathom line off the US coasts. At the same time, the USSR also developed and began retrofitting (by 1973) a third variant of the SS-N-6 that carries MRV warheads. Known as the SS-N-6 Mod 3, its range is also 1600 NM, and the two or three warheads of its reentry vehicle are estimated to have a yield of 350 kilotons each. The SS-N-6 was deployed mainly on the Yankee Class SSBNs.

The SS-NX-17 is a fourth-generation SLBM. The missile is composed of two stages and is the first SLBM to use solid fuel. It has a range of less than 2150 NM and is equipped with a post-boost vehicle characteristic of a MIRV-capable missile – although no MIRV flight tests have been observed. A single modified Yankee Class SSBN (known as Yankee II), with only twelve launch tubes, has been deployed to serve as a testbed for the SS-NX-17. This missile may have been the initial step toward Soviet development of a longer-range solid-fuel SLBM, or possibly a failed attempt to design a higher-performance SLBM to be retrofitted into modified launch tubes of the Yankee SSBNs.

Intercontinental-range Missiles
Intercontinental-range missile development has been one of the most challenging and high-priority areas in Soviet post-war weapon-system work. These weapons are dedicated to covering many different types of targets located primarily in the US mainland. Until recently, Soviet ICBMs were clearly the predominant means for covering such targets, because only small numbers of heavy bombers and sea-based nuclear forces were available for this task through the late 1960s. Since that time the USSR has produced significant numbers of long-range SLBMs capable of striking intercontinental-range targets. Deployed on Delta

Number of Soviet Intercontinental-range Weapons and Warheads, 1960-80

Instrument	1960	1965	1970	1975	1980
Weapons	149	434	1456	1652	1696
Land-based missiles	4	224	1220	1267	1018
SS-6	4	4	0	0	0
SS-7	0	197	197	190	0
SS-8	0	23	23	19	0
SS-9	0	0	240	278	0
SS-11[1]	0	0	720	650	320
SS-13	0	0	40	60	60
SS-17	0	0	0	10	150
SS-18	0	0	0	10	308
SS-19[1]	0	0	0	50	180
Sea-based missiles	0	15	41	196	522
SS-N-4, SS-N-5	0	15[2]	9[2]	0	0
SS-N-6	0	0	32[2]	64[2]	64[2]
SS-N-8	0	0	0	132[3]	282[3]
SS-N-18	0	0	0	0	176
Bombers[4]	145	195	195	189	156
M-4	35	85	85	85	56
Tu-20	110	110	110	104	100
Warheads	294	381	1403	1875	6156
Land-based missiles	4	224[5]	1220[6]	1537[7]	5140
Sea-based missiles	0	15	41	196	876
Bombers	290	142	142	142	142

[1] Includes only those believed to be primarily dedicated to covering intercontinental-range targets.
[2] Forward-deployed missiles.
[3] Including 6 missiles in testbed launchers. [4] Includes tankers.
[5] Plus a reload capability of 142 SS-7s and SS-2s.
[6] Plus a reload capability of 142 SS-7s and SS-5s.
[7] Plus a reload capability of 131 SS-7s and SS-5s.

Class submarines, these sea-based missiles may be (in operational terms at least) merely an extension of the Soviet land-based ICBM force.

Given this predominance of the ballistic missile in the Soviet strategic force posture, the USSR has attempted to instill a basic flexibility into its missile force comparable to that traditionally manifested by the US strategic bomber force. As noted, certain Soviet missiles possess a variable-range capability enabling them to attack targets at regional as well as intercontinental ranges. In some cases, these missiles have merely had their ranges upgraded, while in other missiles this flexibility was part of the original design. Such systems are discussed in this section. Soviet ballistic missiles (particularly ICBM MODs) with the capability to strike targets at global distances are examined in the following section.

The SS-6 Sapwood was the world's first ICBM. It was the culmination of the Soviet Union's post-war research and development effort. Some 40-50 short-range test shots took place during 1956. The SS-6 was first tested at full range in August 1957, and there were seven other tests before the end of the year. Between 1959 and 1961 only four SS-6 ICBMs were deployed, although they remained in the SRF inventory at Plesetsk until 1967. The missile used non-storable liquid fuel, which caused many problems including reaction times as long as twelve hours. Its maximum range was almost 4000 NM. Guidance was by radio command, and therefore highly susceptible to electronic jamming during portions of its flight. Although originally designed to carry a heavy 500-kiloton atomic warhead, thermonuclear technology later allowed it to

Characteristics of Soviet Land-based, Medium-range Missiles
Summary: USSR

Generation and missile	Year design began	First flight test	Propulsion system	Guidance system	Warhead type	Basing mode	Year operation began	Number of warheads	Yield per warhead (megatons)	Accuracy (nautical miles)	Range (nautical miles)	Number of missiles deployed
First generation SS-4 Sandal (medium range)	1949-50	1957	Liquid fuel	Radio command; later, inertial	Single	Soft site, with 4 launchers and ability to refire; hard site, with 4 launchers	1958	1	2.0	1.5	1100	100
SS-5 Skean (intermediate range)	1952-53	1959-60	Liquid	Radio command; later, inertial	Single	Soft site, with 4 launchers and ability to refire; hard site with 3 launchers	1961	1	4.0-6.0	1.0	2200	600
Second generation SS-11 Sego, mod. 1 (variable range	1955-58	1965	Liquid	Fly-the-wire, inertial	Single	Hardened silo	1970	1	0.95	0.76	5900	320
SS-15 Scrooge (intermediate range)	1958-61	1968	Solid fuel	Inertial	Single	Mobile	n.a.	1	0.6	0.5-1.0	3000-4000	n.a.
Third generation SS-19 (variable range)	1966	1973	Liquid	Fly-the-wire, onboard digital computer	Multiple or single	Hardened silo	1975	6	0.55	0.14-0.19	5200-5450	120
SS-20 (intermediate range)	1965-68	1974-75	Solid fuel	Inertial	Multiple	Mobile	1977	3	0.15-0.50	0.16	2700	180

Note: Missile numbers are those used by US military services; names are those used by NATO forces. *Not available.

Characteristics of Soviet Land-based, Intercontinental-range Missiles
Summary: USSR

Generation and missile	Year design began	First flight test	Propulsion system	Guidance system	Warhead type	Launching mode	Base	Year operation began	Number of warheads	Yield per warhead (megatons)	Accuracy (nautical miles)	Range (nautical miles)	Throw weight (pounds)	Number of missiles deployed
First generation SS-6 Sapwood	1949-50	1957	Non-storable liquid fuel	Radio command	Single	n.a.*	Fixed site	1959-61	1	5	2.0	3200	7000-9000	4
Second generation SS-7 Saddler Mod. 1 & 2	1954	1961	Liquid fuel	Radio command	Single	n.a.	Fixed site	1962	1	3.0	1.5	5900	3000-4000	197
Mod. 3		1961						1963	1	6.0	1.0			
SS-8 Sasin	1954	1961	Non-storable liquid fuel	Radio command	Single	n.a.	Fixed site	1963	1	3.0	1.0	5400	2500-	23
Third generation SS-9 Scarp Mod. 1	1957	1964	Liquid fuel	Fly-the-wire, inertial	Single	Hot	Hardened silo	1967	1	20.0	0.5	6500	9000-11000	288
Mod. 2		1964-65			Single			1966	1	20.0	0.5			
Mod. 3		1965			Single			1969	1	20.0	n.a.			
Mod. 4		1968			Multiple			1971	3	3.5	1.0			
SS-11 Sego Mod. 1	1955-58	1965	Liquid fuel	Fly-the-wire, inertial	Single	Hot	Hardened silo	1966	1	0.95	0.76	5900	1000-2000	1030
Mod. 2		1969			Single			1973	1	1.10	0.59	6500		
Mod. 3		1969			Multiple			1973	3	0.35	0.59	5700		
Mod. 4		1974			Multiple			n.a.	3-6	n.a.	n.a.	n.a.		
SS-13 Savage Mod. 1	1958-62	1965-69	Solid fuel	Fly-the-wire, inertial	Single	Hot	Hardened silo	1967-69	1	0.6	1.0	5075	1000	60
Mod. 2		1970			Single			1972	1	0.6	0.82			
Fourth generation SS-X-16	1965	1972	Solid fuel	Fly-the-wire, on-board digital computer	Single	Hot	Mobile and hardened silo	...	1	0.65	0.26	4970	2000	...
SS-17 Mod. 1	1965	1972	Liquid fuel	As above	Multiple	Cold	Hardened silo	1975	4	0.75	0.24	5400	8000	150
Mod. 2		1976			Single			1977	1	3.6	0.23	5900		
Mod. 3		n.a.			Multiple			1979	4	0.75	n.a.	n.a.		
SS-18 Mod. 1	1965	1972	Liquid fuel	As above	Single	Cold	Hardened silo	1974	1	24.0	0.23	6500	16000	308
Mod. 2		n.a.			Multiple			1976	8-10	0.9-0.55	0.23	5900		
Mod. 3		n.a.			Single			1976	1	20.0	0.19	8640		
Mod. 4		n.a.			Multiple			n.a.	10	0.55	0.14	5400		
SS-19 Mod. 1	1966	1973	Liquid fuel	As above	Multiple	Hot	Hardened silo	1975	6	0.55	0.19	5200	6000	300**
Mod. 2		n.a.			Single			1978	1	4.3	0.21	5450		
Mod. 3		n.a.			Multiple			1979	6	0.55	0.14	5200		

Note: Missile numbers are those used by US Military Services; names are those used by NATO forces. *Not available. **Another 60 in preparation.

Characteristics of Soviet Global-range Missiles

Missile	Year design began	First flight test	Propulsion system	Guidance system	Warhead type	Basing mode	Year operation began	Number of warheads	Yield, per warhead (megatons)	Accuracy (nautical miles)	Range (Nautical miles)	Throw weight (pounds)	Number of missiles deployed
SS-9 Scarp, Mod. 3	n.a.	1965	Liquid fuel	Fly-the-wire, inertial	Single	Hardened silo	1969	1	20.0	1.0-2.0 / 1.5-3.0	Depressed trajectory / FOBS*	9000-	10
SS-X-10 Scrag	1957-58	1964-65	Non-storable liquid fuel	Fly-the-wire, inertial	Single	n.a.	...	1	20.0	1.0-2.0 / 1.5-3.0	Depressed trajectory / FOBS*	9000-11000	...
SS-18. Mod. 3	n.a.	1975	Liquid fuel	Fly-the-wire, onboard digital computer	Single	Hardened silo	1976	1	20.0	0.19	8600	16000	n.a.

*Fractional orbital bombardment system

Characteristics of Sea-based Missiles

Generation and missile	Year design begun	First flight test	Propulsion system	Guidance system	Warhead type	Submarine assignment	Year operation began	Number of warheads	Yield, per warhead (megatons)	Accuracy (nautical miles)	Range (nautical miles)
First generation SS-N-4 Sark	1949-50	n.a.	Liquid fuel	Inertial	Single	Golf I: Hotel I	1958	1	2-0-3.5	2.0	350
Second generation SS-N-5 Serb	1954-55	n.a.	Liquid fuel	Inertial	Single	Golf II; Hotel II	1963	I	4.0	1.5	750
Third generation SS-N-6 Sawfly											
Mod. 1	1960	1967	Liquid fuel	Inertial	Single	Yankee I; Golf IV for testing	1968	I	0.7	1.0	1300
Mod. 2	1960	1972	Liquid fuel	Inertial	Single	Yankee I; Golf IV for testing	1973	I	0.65	1.0	1600
Mod. 3	1960	1973	Liquid fuel	Inertial	Multiple	Yankee I; Golf IV for testing	1973	2-3	0.35	1.0	1600
Fourth generation SS-N-8											
Mod. 1	1962	1969	Liquid fuel	Stellar inertial	Single	Delta I and II: Hotel III; Golf III	1973	I	n.a.	n.a.	4200
Mod. 2	1962	1976	Liquid fuel	Stellar inertial	Single	Delta I and II: Hotel III; Golf III	n.a.	I	0.8	0.84	4900
Fifth generation SS-NX-17	1969	1976	Solid fuel	n.a.	Single, post boost vehicle	Yankee II for testing	n.a.	I	n.a.	n.a.	2100
SS-N-18											
Mod. 1	1969	1976	Liquid fuel	Stellar inertial	Multiple, independently targeted	Delta III	1978	3	0.2	0.76	3500
Mod. 2	1969	n.a.	Liquid fuel	Stellar inertial	Single	Delta III	n.a.	1	0.45	0.76	4300
Mod. 3	1969	n.a.	Liquid fuel	Stellar inertial	Multiple, independently targeted	Delta III	n.a.	7	n.a.	n.a.	3500
Sixth generation SS-NX-20	1973	1980	Solid fuel	n.a.	Multiple independently targeted	Typhoon	n.a.	12	n.a.	n.a.	4500

Note: Missile numbers are those used by US military services; names are those used by NATO forces.
*Not available.

Characteristics of Soviet Air-to-Surface Missiles

Missile	Range	Altitude	Year Operation Began	Weight at Liftoff	Number Deployed
SA-1 Guild	25 NM	Medium	1954	–	3500 Rails
SA-2 Guideline	25 NM	Medium	1959	–	Phasing out
SA-3 Goa	15 NM	Low	1961	–	400 Sites
SA-5 Gammon	175 NM	High	1963	–	100 Complexes
SA-10	55 NM	Low	1982	–	?
Griffon ABM	–	High	Canceled	8 Tons	–
Galosh ABM	400 BM	High	1968	36 Tons	64
SW-8 ABM	–	Low	?	4 Tons	–

carry a five-megaton warhead. The SS-6 ICBMs were based in above-ground launch facilities. Despite its limited success as an ICBM, the SS-6 eventually became the workhorse of the Soviet Union's space program as the basis of its booster rockets.

The SS-7 Saddler was a second-generation ICBM operated by the SRF. This was a follow-on design to the SS-6, which underwent full-range flight testing in the Pacific during 1961 and was the first Soviet ICBM to be mass produced, beginning in 1962. Initially using a radio-command guidance system, it was a two-stage missile and was the first Soviet ICBM to employ storable liquid fuel.

There have been three models of the SS-7, all with a range of 6000 NM. The SS-7's Mods 1 and 2 had a 3.0-megaton warhead, while the Mod 3 was reported to have a 6.0-megaton warhead. One hundred and twenty-eight soft launchers were in service by late 1965, with one reload missile per launcher to meet reliability and retargeting goals. Another 69 launchers were in underground silos for greater protection against US ICBM strikes.

The SS-8 Sasin was also a second-generation ICBM. It was developed as a direct improvement of the SS-6 and was deployed in 1963. The SS-8 had two stages and utilized unstorable liquid fuel. Guidance was by radio command, while its 6500-NM range expanded the targeting flexibility of the Soviet ICBM force. Like the SS-7 Mod 3, its warhead was 6.0 megatons, although at one time it was thought to have emphasized electromagnetic pulse (EMP) effects rather than blast. Only 23 SS-8 launchers were deployed. Nine were in underground silos, while fourteen others were deployed in soft above-ground coffins.

The SS-9 Scarp was a third-generation ICBM operated by the SRF and based on the SS-7 experience. The SS-9 was first flight tested in late 1963. Two versions became operational – during 1966, Mod 2, and in 1967, Mod 1. The missile used storable liquid fuel and incorporated a fly-the-wire guidance system.

There are five modifications of the SS-9 missile frame. Mod 1 had a range of 4300 NM and a warhead yield of 12 megatons. Mod 3 is a Fractional Orbital Bombardment System (FOBS).

The SS-9 Mod 4 was the first Soviet multiple-reentry-vehicle (MRV) system. It had a range of 6500 NM and carried three unguided 3.5-megaton warheads, which were dispatched from a rail system at regular intervals. It was first tested in August 1968 and continued testing through 1970. The SS-9 Mod 5 is not located at regular SRF bases. Instead, as a space booster for the Soviet anti-satellite systems, it is launched from above-ground pads in Tyuratam. Altogether, 288 SS-9 missiles were deployed in underground launch silos. The majority of these silos contained SS-9 Mod 2s and some Mod 1s, while some Mod 4s may also have been deployed.

The SS-11 Sego is a third-generation strategic missile notable for its variable-range capability, which enables it to strike targets at regional as well as intercontinental ranges. It was the first 'light' ICBM developed by the USSR and as such reflected a change in Soviet targeting requirements. The missile was first tested in 1965 and was deployed in 1966. Composed of two stages, the SS-11 uses storable liquid fuel and has a fly-the-wire guidance system.

There are known to be four different models of the SS-11. The Mod 1 may have had initially a range of only 3000 NM, commensurate with its origins as a naval or regional strike system. Later, its range (and probably its firing parameters) were improved. It carried a 950-kiloton warhead. The Mod 2 was first tested in 1969 and became operational in 1973. While its external frame is similar to that of the SS-11 Mod 1 in all respects, it was the first Soviet ICBM to carry such penetration aids as chaff or decoys. The Mod 2 had a range of 5600 NM and a warhead yield of about 1 megaton. The Mod 3's testing and deployment were similar to those of the Mod 2. As the first operational MRV system in the Soviet inventory, this missile has a maximum range of 5600 NM and carries three 350-kiloton warheads. The MRV warheads operate on a rail dispensing system rather than a post-boost vehicle reentry basis. Finally, the SS-11 Mod 4 began testing in the mid-1970s. It also has an MRV capability, but one different from that of the Mod 3; the Mod 4's warheads reportedly reenter the atmosphere simultaneously and at a very low speed. This could make it applicable for such specialized missions as chemical strikes against aircraft carriers or possibly even for delivering ASW weapons. All told, some 1030 SS-11 launchers were constructed in modestly hard underground silos, including over 300 launchers in missile fields near Europe and China.

The SS-13 Savage is also a third-generation ICBM, which represented a shift away from large missiles. The SS-13 was first tested in 1965 and was deployed in 1967. The missile has three stages and was the first Soviet ICBM to use solid fuel. The SS-13 used fly-the-wire guidance, has a range of 5600 NM and carried a single warhead of 600 kilotons. Testing of a Mod 2 version of the SS-13 began in the early 1970s. This version had fins for greater stability at launch. Only 60 SS-13s were ever deployed, and these were located in hard underground silos. The small number deployed is probably due to the problems encountered with the SS-13's last-stage solid-fuel rocket motor. At one time the US had estimated that up to 300 SS-13s might be deployed, with 100 on mobile launchers. This would have filled a pressing Soviet requirement for a survivable strategic reserve.

The SS-X-16 was a fourth-generation ICBM. A three-stage, solid-fuel missile, it was a direct follow-on to the SS-13. It was first flight tested in March 1972, but never deployed due to limitations in the proposed SALT II Treaty. The SS-X-16 incorporated an onboard digital computer with a traditional fly-the-wire guidance system and a post-boost vehicle for improved accuracy. It had a range of 5000 NM and could carry a warhead of 650 kilotons.

The SS-17 is a fourth-generation ICBM produced as a replacement for the early model SS-11. It was first tested in September 1972 and became operational in 1975. The SS-17 is a two-stage, storable liquid-fueled missile. It also has a 'cold launch' capability that ejects the missile from the silo before engine ignition takes place. Its guidance is similar to that of the SS-X-16. There are two versions of the SS-17: the Mod 1 has a range of 5400 NM and a MIRV'd post-boost reentry vehicle with four 750-kiloton warheads. The Mod 2 was first tested in February 1976 and has a longer range of 5800 NM with a single 3.6-megaton warhead. Another version of the SS-17, the Mod 3, may also serve as a testbed for a new ICBM development. Some 150 SS-17s have been deployed in refurbished and much harder underground silos than those of the SS-11s. Deployment of the SS-17 is probably also intended to take over the targeting responsibilities once held by the earlier SS-7 ICBMs.

The SS-18 is a fourth-generation ballistic missile designed as a follow-on to and replacement for SS-9. Its first test flight occurred in December 1972. By November 1974 it had been tested 29 times with seven failures. It first became operational in 1974. The missile incorporates storable liquid fuel, while its guidance package is similar to that found on the SS-X-16.

There are presently five versions of the SS-18. The Mod 1 has a range of 6500 NM and a warhead yield of 24 megatons. The Mod 2 has a range of 6000 NM and a MIRV'd warhead with two combinations, the first having eight warheads with yields of 900 kilotons each, and the second having ten warheads, each rated at 550 kilotons. The Mod 2 has had development problems. The Mod 3 has a range of 8700 NM and carries a single lightweight 20-megaton-yield warhead. The Mod 4 has a range of 6000 NM and carries 10 500-kiloton warheads. It represents a refinement in accuracy and will replace the Mod 2. The Mod 5 may be a future testbed incorporating some type of maneuvering guidance system. Like the SS-17, the SS-18 is a 'cold launch' missile with rocket ignition occurring outside the silo. The USSR has about 308 launchers constructed in hard underground missile silos capable of accepting the SS-18.

The SS-19 is also a fourth-generation ICBM, a direct follow-on in design and mission to the SS-11. First tested in April of 1973, it became operational in December 1974 after very successful flight testing in which reliability values of 90 percent were reportedly achieved. The SS-19 is composed of two stages and uses storable liquid fuels. Its guidance package is similar to that of the SS-X-16.

There are three known versions of the SS-19. The Mod 1 has a range of 5200 NM and carries six MIRV'd 550-kiloton warheads dispensed from a post-boost vehicle. The Mod 2 began flight testing in 1976. It has a range of 5500 NM with a single 4.3-megaton warhead. The SS-19 Mod 3 is similar to Mod 1, but with improved range and accuracy. Some 360 SS-19s have been, or are about to be, deployed, including 180 in medium-range fields for regional strike missions. Almost all are located in very hard underground silos formerly occupied by the SS-11.

Soviet intercontinental-range SLBMs appear to be distinguished from their shorter-range predecessors by their greater dedication to the role of intercontinental strategic strike. Their many similarities to land-based ICBMs suggest that in wartime they would be directly

controlled by the Supreme High Command. Compared to its land-based counterparts, sea-based intercontinental missile development is a relatively recent phenomenon.

The first intercontinental-range Soviet SLBM was the SS-N-8, a fourth generation SLBM. A major improvement of the SS-N-6 design, it was first tested in 1969 and went to sea in 1972. The SS-N-8 carries a single 800-kiloton warhead. The missile has two main stages and uses storable liquid fuels. Guidance is fly-the-wire, but in order to maintain its accuracy at this great distance, the SS-N-8 has a stellar inertial guidance system that updated the missile's flight path by utilizing two 'star fixes' for mid-course corrections. The initial range of the SS-N-8 was some 3000 NM, but during its long development period its range was upgraded to 5400 NM through an increase in missile volume as well as improved fuels. Its increased range and its deployment date were consistent with the SS-N-8's new strategic reserve mission, which would include deep strikes upon US mainland targets. About 282 SS-N-8 launchers have been deployed on Delta I- and Delta II-Class SSBNs.

A direct product improvement of the SS-N-8, the SS-N-18 is also a fourth-generation SLBM. It was first tested in 1976, and after an extensive series of tests in 1977 became operational during 1978. The missile is similar in size and shape to the SS-N-8. It also uses storable liquid fuels and has a more sophisticated guidance system than the SS-N-8.

There are several SS-N-18 models. Mod 1 has a range

Above: The SS-14 is a solid-propellant MRBM made up of the upper two stages of the SS-13 ICBM. It is pictured on its transporter in a Red Square parade, but is fired from a fixed site.

of 4600 NM, and it is noteworthy as the first Soviet SLBM to have been tested with MIRV warheads. Mod 2 has three 200-kiloton warheads, while the Mod 3 carries seven reentry vehicles, although over a much shorter range. The Mod 2 has a single 450-kiloton warhead. About 192 SS-N-18 launchers were deployed by 1980 on the Delta III SSBN, which carries 16 SLBM launchers each. The long range of this missile, coupled with its capability to deliver MIRV warheads, will significantly enhance Soviet sea-based coverage of targets deep in the US homeland by Soviet SLBMs.

The SS-NX-20 is a fifth-generation SLBM. A larger missile, it appears to be a follow-on design to both the SS-N-8 and SS-N-18. Flight testing of the SS-NX-20 began in early 1980 and has been plagued by a series of failures. With its use of solid-fuel propellant, this SLBM seems aimed at combining a long-range capability with the guidance and payload benefits of a MIRVed SLBM. If successfully developed and deployed aboard the Typhoon SSBN, this SLBM would be a major improvement to the Soviet Union's sea-based strategic reserve capability.

Global Missiles

Similar to ICBMs, global missiles are driven primarily by the broad targeting requirements placed on the Soviet ICBM force. Due to their large size, they have also been readily adaptable to changes in targeting doctrine as well as to supporting the Soviet space program. In addition to covering such area targets as SAC airbases or economic centers, global missiles also have some interesting military roles outside the range of ICBMs.

An earlier global missile was the Proton 'space booster' that began design in 1958. The first version flew in 1965 and a second was launched in 1967. In a military configuration (designated SS-XL Scarp) the Proton Mod 1 had the capacity to deliver a 35-45-megaton-yield warhead while the Proton Mod 2 could have carried a 45-55-megaton-yield warhead. At the time, this would have continued the Soviet trend toward increasingly larger warheads in the SRF inventory to meet its area targeting requirements. In contrast, the *G*-class space booster flight tested in 1969 was clearly built without the constraints of military development programs. Its design work began about the same time the Soviets were engaged in a determined space program with the Cosmos series. As a result, its enormous payload was not a continuation in the trend lines of dual-use rockets that had begun with the SS-6 and terminated with the Proton.

One model of the SS-9 Scarp was developed with a global strike capability. The SS-9 Mod 3 was actually a Fractional Orbital Bombardment System (FOBS), and therefore capable of attacking around the South Pole, or alternatively, of depressed-trajectory shots over the North Pole. It carried a 10-megaton warhead. The first SS-9 Mod 3 test took place in 1965, and through 1969 there were 20 flight tests, with about 75 percent considered successful. One test in 1969 was a crew-training shot, suggesting that an operational capability had been obtained. Testing terminated in August 1971. Some 18 Soviet launchers were apparently dedicated to FOBS operations.

The SS-10 Scrag was another third-generation global missile, a follow-on design to the SS-8. Although successful testing took place from 1964 to 1966, it was never deployed, probably due to political and military reasons. A large missile, the SS-10 had three stages and was the last unstorable-liquid-fueled missile tested by the USSR. Its guidance was fly-the-wire, and its full range may have been better than 6500 NM with a 20-megaton warhead. The SS-10 was also the first to use variable nozzles in the first of its three stages, and thus had FOBS capability prior to the SS-9 Mod 3, although it was never tested as such.

In the latest generation of ICBMs, only the 8700-NM range of the single-warhead SS-18 Mod 3 approaches the range of a global missile. Crew-training launches of the Mod 3 began in February 1976. The range of the SS-18 allows coverage of global areas that contain potential adversaries of the USSR. These might include countries with an interest in or capability of producing nuclear weapons, such as certain South American countries or South Africa. The SS-18 Mod 5 also has sufficient range to strike US naval forces almost anywhere on the high seas. Finally, the USSR may also see its new long-range SLBMs (discussed earlier) as a feasible means for attaining global missile coverage.

Soviet Missile Deployments

While great attention has been focused on the evolution of Soviet missile effectiveness, much less has been accorded to an equally important development, trends in Soviet missile deployments. Yet over the past 20 years steady and impressive advances have occurred in the Strategic Rocket Forces' missile basing system as well as in Soviet sea-based missile deployments. These improvements have been a prerequisite to enhancing the overall effectiveness of the Soviet strategic forces. Furthermore, the nature of Soviet missile deployments has reflected an evolving process characterized by a distinctive set of priorities and concerns.

Land-based Strategic Missile Deployments

Soviet land-based missiles have been deployed in four basic configurations. These include above-ground unprotected launching pads, above-ground protected sites, hardened underground silos and land-mobile systems. Over the past two decades Soviet missile deployments have become increasingly survivable, either through harder and better-constructed fixed sites, or through the deployment of missiles with a mobile capability. To some degree, this gain in force survivability required the USSR to sacrifice some of the reload capability that was a notable feature of its earlier missile deployments.

The first Soviet strategic missiles, the regional-range SS-4 and SS-5, were initially deployed in clusters of relatively soft above-ground launchers. This type of missile-basing system was used through the early 1960s. The individual launch sites comprised four separate missile launchers, each with a reload missile.

Above: SS-15 Scrooge is a giant mobile weapon system, thought to be the largest such ever put into service. Each missile tube rides on an IS-3 tank chassis. This system is also called 'XZ'.

About one-fifth of the nearly 650 regional missiles were later deployed in hardened shelters, with three or four launchers in a cluster. From 1963 to 1966, 135 SS-4 and SS-5 missiles were placed in hardened shelters, and then the hardening program ceased. A number of factors may account for this, including the more immediate priority of the SS-9 and SS-11 silo-construction program which began in 1964. Another may have been a Soviet desire to retain the missile-reload capability associated with the softer sites.

One result of this cessation was that the regional missiles of the Strategic Rocket Forces continued to be relatively vulnerable to attack by Western nuclear forces. The USSR may have planned to alleviate this problem by deploying mobile missiles as replacements for the SS-4 and SS-5 force. Soviet tacticians of the 1960s frequently praised the greater survivability of mobile missiles due to their ability to be concealed from enemy surveillance. In the mid-1960s the Soviet Union started testing two missiles, the SS-14 MRBM and the SS-15 IRBM, which employed mobile launch systems. However, due to problems with these solid-fuel missiles, neither was deployed in large numbers for any length of time.

During the 1960s the USSR also undertook programs to enhance the survivability of its growing intercontinental-range missile force. Here Soviet efforts were relatively more successful, although US improvements to its Minuteman ICBMs continued to keep a large portion of the SRF's missile force vulnerable for some time. But by the end of the decade, the overall survivability of the Soviet intercontinental missile force was substantially upgraded.

The early Soviet ICBM deployments were relatively vulnerable by any standards. The initial Soviet ICBM, the SS-6, was deployed at soft above-ground launch pads at Plesetsk. Both sites had exposed launch areas and support facilities due to the awkwardness and volatility of the SS-6's unstorable liquid fuel. As the second generation SS-7 and SS-8 were deployed, greater attention was given to both their protection and their concealment. Nonetheless, like the SS-4 and SS-5 force, most of the SS-7s and SS-8s were deployed in soft concentrated clusters with two launchers per site.

One advantage of basing missile launchers in a cluster was the ease of reloading the launchers once they had been fired. Another consideration was that clustered missile launchers were probably easier to defend with SAMs against enemy bomber attack. But by the time these missiles became operational in the early 1960s, the major US threat was shifting from bombers to missiles. With ICBMs, however, warning time was minimal and defenses did not exist. Even a nominal one-megaton blast delivered by an ICBM detonating as far away as one-half mile would have an excellent chance of disabling the entire cluster. Obviously, the reload advantage was negligible if the launchers were destroyed. Consequently, beginning in 1962, the Strategic Rocket Forces sought to provide greater protection to its ICBMs by placing them in separate coffin-like revetments, with three launchers per site. Later, a certain proportion would be placed in first-generation underground silos. To some degree these early measures reflected the priority that the USSR placed on such combat qualities as survivability and reload – priorities that would be affirmed in succeeding generations.

With the deployment of the third-generation ICBMs, the SS-9 and the SS-11, the clustering of launchers at a single missile site ceased. All SS-9, SS-11 and SS-13 ICBMs were deployed in separated underground silos, hardened up to about 350 psi against blast effects. The deployment of these hardened silos also reflected the USSR's shift in priorities to emphasize missile protection over providing for a rapid refire capability.

It was the fourth-generation ICBM deployments that resulted in major strides to Soviet ICBM force survivability. All of the new SS-17, SS-18, SS-19 and SS-11 Mod ICBMs were emplaced in very hard silos estimated to be capable of withstanding up to 4000 psi in blast overpressure. Part of this upgrade resulted from a tremendous amount of concrete, as well as internal cushioning to insure the missile's survival. In addition to making the silos harder, survivability was further enhanced by removing missile support equipment from around the lip of the silo, where it was highly vulnerable to attack, and placing it inside the silo itself. Once again, Soviet interest in retaining some form of reload capability resurfaced with the development of certain 'cold-launched' ICBMs like the SS-17 and SS-18. By withholding engine ignition until after the ICBM is propelled from the silos, the missile silos can be more easily refurbished and reloaded with reserve missiles.

The fourth-generation ICBM deployments provided the USSR with its greatest confidence to date in the ability of its ICBM force to generally ride out or survive an enemy counterforce attack. At least for the near future, the US will lack a high-confidence capability, in terms of large numbers of very accurate warheads, to effectively neutralize the hardened Soviet missile force.

The original Soviet motivation for undertaking these and other expensive hedges against ICBM vulnerability could have been fostered by US interest in deploying a major new ICBM by the late 1970s. During the late 1960s the US reportedly considered developing a much more accurate, large throw-weight ICBM known as the WS-120A, which was scheduled for deployment within the decade. The US later deferred initiation of development work on a new, more powerful ICBM for some years until the current M-X Peacekeeper ICBM program. Nonetheless, possible Soviet Union expectations that such a weapon would be deployed during the lifetime of its fourth-generation ICBMs may have convinced it to devote substantial resources to ensuring its missiles' survivability.

Also important among the missile survivability measures the USSR pursued in the 1970s was the development of mobile missile launchers. Because of the difficulty of finding and attacking mobile missile launchers, these systems offer one possible means for fulfilling the Soviet requirement for a survivable strategic reserve force. As noted earlier, Soviet interest in mobile strategic missiles was initially manifested in the mid-1960s with the SS-14 and SS-15 mobile systems. At the time, Western analysts also expected the solid-fuel SS-13 ICBM to have a mobile capability. Due to technical problems, however, it was only deployed in small numbers in silos.

In the 1970s the USSR pursued development of two new solid-fuel mobile missiles: the SS-20 IRBM and the SS-X-16 ICBM. Both mobile systems may have been motivated in part as another hedge against an advanced US ICBM in the 1970s. Of the two missiles, only the SS-20 is currently operational because of Soviet adherence to the proposed SALT II Treaty not to deploy the SS-X-16 in order to avoid verification problems with the technically similar SS-20.

The SS-20, which is now replacing the older SS-4/SS-5 force, is the SRF's first operational mobile missile. It is not likely, however, to travel randomly in search of concealed firing areas. More probably, a regiment of SS-20s will be garrisoned at a central base and will disperse periodically to local pre-surveyed areas to await firing instructions. This would afford greater protection than the SS-4 and SS-5 forces, since the dispersed missiles would increase their survivability while still enjoying air defense cover, maintaining economy in operation and allowing a reload capability to be kept in the garrison area.

Finally, the next generation of Soviet land-based missiles, to be deployed in the 1980s, is likely to continue the trends of both greater hardening and more mobility. The new silo-based ICBM could be deployed in silos hardened up to 6000 psi. Whether this addi-

Left: Whisky Long Bin was the NATO name for a group of diesel submarines rebuilt to fire four N-3 cruise missiles.

Below: In addition to ballistic missiles far more numerous than those of all other countries combined, the Soviet Union is also by far the biggest user of long-range cruise weapons. This Juliet-class submarine has two pairs of large N-3 Shaddock missiles.

Far right: The Echo II submarines each carry four twin N-3 launchers.

tional investment in hardness will be able to offset US advances in missile accuracy is open to question. Indeed, once certain accuracy thresholds have been crossed (at about 600 feet) Soviet confidence in silo hardness as a means of riding out an ICBM attack could become fairly low. More likely, the USSR will place even greater emphasis on the successful development of a new mobile missile in this next generation of ICBMs.

Sea-based Strategic Missile Deployments

Although relying on somewhat different means, the sea-based strategic missile force has also achieved great gains in effectiveness and survivability over the past twenty years. Unlike land-based ICBMs, missile range and factors other than the submarine launch platform itself have been critically important in enhancing Soviet SLBM survivability. Nonetheless, the growth of the Soviet sea-based ballistic-missile force, in numbers as well as presence, has been reflected in the evolution of the SLBM's submarine launch platform. This section will review the development of the Soviet ballistic-missile submarine force and its operating practices.

In the early post-war years the USSR explored the feasibility of previous German experiments toward developing a submarine-launched ballistic missile using sea-going containers towed by submarines as launch platforms. By the early 1950s Soviet efforts shifted to developing a missile that could be launched directly from the submarine. The USSR completed development of its first-generation sea-launched ballistic missile, known as the SS-N-4 Sark, in 1957 and deployed it upon six converted Soviet patrol submarines. These submarines, known as the Z-V Class, were medium-range diesel-powered submarines of the Zulu Class reconfigured to carry two SS-N-4 missile launchers in their conning towers. The primary disadvantages of this early system were that the missile could be launched only while the submarine was exposed on the surface and that the SS-N-4 had a relatively limited range of 350 NM. Regardless, the handful of Z-V Class SSBs deployed provided the USSR with an initial sea-based ballistic missile force, as well as a prototype system for crew training and further development.

The Z-V Class was soon followed by the deployment of two new classes of submarines: the Golf Class SSB and the Hotel Class SSBN, both deployed between 1958 and 1962. The diesel-powered Golf Class submarine carries three ballistic-missile launch tubes in its conning tower. A total of 23 units of this class was originally deployed, although one Golf Class SSB sank in the Pacific in 1968. At about the same time, the USSR deployed the Hotel Class SSBNs, its first nuclear-powered ballistic-missile submarines. Only eight submarines of the Hotel Class were deployed. Like the G Class, this submarine carried three missile launch tubes in its conning tower, and both were originally deployed with the surface-launched SS-N-4 SLBM.

This major shortcoming of the Golf and Hotel Class submarines was corrected beginning in 1963 when the more advanced SS-N-5 SLBM, which could be launched from a submerged submarine, became operational. The SS-N-5 was retrofitted upon most of the H Class SSBNs between 1963 and 1967 and subsequently upon 13 of the G Class SSBNs following 1967. These submarines became known as the Hotel II and Golf II Classes, as opposed to the Golf I Class that still carried the earlier SS-N-4 SLBM. Later, a few of the SS-N-8 and SS-N-6 SLBMs were backfitted on these submarines, causing their nomenclature to be changed to Golf III, Golf IV and Hotel III submarines.

In 1967 the third-generation Soviet SSBN, the Yankee Class, went to sea and became operational the next year. This submarine was a significant improvement over its predecessors. It carried several more ballistic missile launch tubes and 16 SLBM launchers per submarine and was equipped with a longer-range missile, the 1300-NM SS-N-6.

Some analysts contend that the SS-N-6 was actually used as a 'quick fix' SLBM for the Y Class and that the tactical SS-NX-13 SLBM (tested, but never deployed) was originally intended to be deployed in the Yankee Class SSBN. Between 1966 and 1974, 34 Yankee Class SSBNs were deployed. In recent years the USSR has begun to dismantle a number of Y Class SSBNs as newer Delta Class SSBNs become available, in order to abide by the SALT I agreement limits. One Yankee Class SSBN (Y-II Class) has also been converted into a 12-launch-tube testbed platform for testing the solid-fuel SS-NX-17 SLBM.

By the early 1970s the Yankee Class production run began to wind down, as the USSR initiated deployment of a new modified class of Yankee submarines that was later designated the Delta Class by the West. The initial submarines in this class were deployed with the long-range SS-N-8 SLBM. The first Delta Class SSBN armed with the deep-strike SS-N-8 SLBM became operational in July 1973. This SSBN mounted 12 launch tubes and was part of the first series of the new class of Delta SSBNs that became operational between 1973 and 1976 (the Delta I Class). Deployment of the Delta II series of SSBNs began in 1976. These submarines differed from the Delta Class only in terms of their greater length, which enabled them to accommodate 16 SS-N-8 launchers. The greater range of the SS-N-8 SLBMs deployed on the Delta Class SSBNs enhances their survivability by allowing them to cover their targets in the US from safer home waters, thereby minimizing their exposure to Western ASW forces.

Deployment of the SS-N-8 SLBM was followed by the MIRVed SS-N-18, which became operational in 1978. It is deployed on the Delta III SSBN which carries 16 SLBM launchers. Currently, the USSR has launched a new large SSBN known as the Typhoon Class. It has 20 SLBM launch tubes that are probably intended to carry the new SS-NX-20 SLBM, if it is successfully tested and deployed.

Also indicative of the growing capability of the Soviet ballistic-missile submarine force has been its gradually expanding deployments. Soviet ballistic-missile submarines are currently deployed with three of the four main fleets of the Soviet Navy. About 70 percent of the SSBNs are deployed with the Northern Fleet, based at Murmansk on the Kola Peninsula. The ballistic-missile submarines stationed there are deployed to regular patrol areas in the Barents, Norwegian and Greenland Seas, as well as forward to stations off the eastern coast of the United States. Most of the other Soviet ballistic-missile submarines are stationed with the Pacific Fleet at the submarine base at Petropavlovsk on the Kamchatka Peninsula. Finally, six Golf II SSBs were deployed to the Baltic Fleet for the first time in 1976.

Soviet ballistic-missile submarine deployments have gradually evolved over the past 20 years in terms of their

Left above: A Charlie-class cruise-missile SSGN photographed in 1974.
Above: The awesome 'missile farm' of the impressive new multirole surface combatant *Kirov*.

numbers and patrol areas. The early-generation submarines, the Z-V, G and H Classes, were probably not deployed beyond Soviet home waters until the mid-1960s. After an extensive ocean survey program, the USSR began to patrol Golf and Hotel Class submarines regularly in the open ocean areas by 1966.

In early 1966 the USSR undertook its first underwater around-the-world cruise by nuclear-powered submarines. The regular patrol areas of the G and H Class submarines at this time were west of the Azores and east of Nova Scotia in the Atlantic and west of Hawaii in the Pacific. Although the first Yankee Class SSBN entered sea trials in late 1967, it was not until late 1969 that Y Class submarines initiated continuous patrols along the Atlantic coast of the United States. After 1969 the USSR maintained two Y Class SSBNs on station within range of the US eastern coastline, with one submarine deployed north of Bermuda and the other south of Bermuda. In that period a Hotel Class SSBN continued to be maintained east of Nova Scotia and a Golf Class SSB west of the Azores. Yankee Class submarines began continuous patrols in the northeast Pacific by 1970, with one submarine stationed west of Hawaii.

By 1974 the Soviet Union initiated regular patrols by its new Delta Class SSBNs, which carried the long-range SS-N-8 or SS-N-18 SLBMs. Such long-range SLBMs gave the USSR the option to patrol these submarines close to home waters like the Barents and Greenland Seas, as well as in the northern Pacific. With their very long-range SLBMs, the Delta Class SSBNs can cover most US targets as soon as they depart from their home bases. The USSR presently maintains routine Yankee Class SSBN patrols off the east and west coasts of the US, while the Delta Class SSBNs are deployed closer to home waters. Delta Class SSBNs routinely deploy in the Pacific, as well as to the Norwegian, Greenland and Barent Seas in the Atlantic.

Soviet Air Defenses

One of the greatest distinctions between US and Soviet strategic forces is the priority the Soviets accord to defensive forces. The Soviet Union maintains a large and expensive air-defense system despite the fact that defenses against ballistic missiles are prohibited by the Anti-ballistic Missile (ABM) Treaty. It also continues to spend large sums on other defensive programs, including ABM research and development as well as civilian defense measures. By contrast, the United States allowed its strategic air defenses to decline after the mid-1960s, when ballistic missiles replaced long-range bombers as the major strategic offensive weapon. The United States has also increasingly favored an offense-dominated strategic force posture as the best means of deterrence.

The Soviet Union, however, has generally maintained and expanded its strategic defenses since the mid-1960s. Its greater commitment to strategic defenses arises in part from the fact that strategic bombers continue to be a significant portion of the American strategic force, and bombers and fighter-bombers are the primary attack instruments of its regional nuclear forces. Moreover, the character of Soviet doctrine compels the use of defensive as well as offensive forces to prohibit nuclear attack on the USSR.

Strategic defensive forces would attempt to prevent Western retaliatory forces from attacking Soviet political and military command posts, industrial centers, military complexes and a variety of other targets. The high costs that will result from even a small number of nuclear-armed enemy forces reaching their targets make this an almost impossible mission. Yet given the Soviet Union's conception of the likely duration of a modern war, even a marginal defense capability against the enemy's offensive forces is worthwhile. Furthermore, military establishments often attempt to do the best they can in solving any particular problem while hoping for some further breakthrough in military technology to improve the situation.

All of the Soviet strategic defensive forces were under the command of the National Air Defense Force of the Homeland (PVO Strany) from 1954 to 1980. Within PVO Strany's domain were early-warning radars and satellites, aviation units (APVO), surface-to-air mis-

siles (Zenith Missiles Troops – ZMT), anti-ballistic missiles (PRO Strany) and anti-satellite systems (PKO Strany). Anti-aircraft forces would counter the manned aircraft that both the United States and China rely on to deliver nuclear weapons; anti-missile forces, relegated to one side by the ABM treaty, shield the national command in Moscow from ballistic missile attacks and anti-space forces can reduce the effectiveness of supporting elements of the West's strategic forces.

Air Defense Deployment
In a major national effort to modernize its air defense force – about 1000 fighters of World War II vintage supported by a visual reporting system – the Soviet Union had by 1953 doubled the number of its fighters, and all were of modern jet design. Large numbers of anti-aircraft artillery supplemented the fighter force, and electronic early-warning systems allowed detection of US and British bombers before they reached the Soviet homeland. Two special National Air Defense districts were created with headquarters in Moscow and Baku to cover what has been traditionally considered the political and economic heartland of the USSR, and the Warsaw Pact countries were brought into the Soviet strategic air-defense system. Relying primarily on the early-model interceptors, the USSR was equipped to counter the kind of high-altitude, clear-weather bombing raids in daylight that the US B-29s had carried out over Korea. But by this time the West had shifted its planning to an air offensive employing the more advanced B-36 and B-47 strategic bombers, whose operations were limited neither by weather nor by altitude. Setting a pattern that would be replayed time and again, the large Soviet air-defense system seemed to trail one step behind the strategic bomber threat it was to counter.

During the mid-1950s concentric circles of early SA-1 Guild launchers were deployed around Moscow for the protection of the top Soviet leadership. Shortly afterwards, the SA-2 Guideline, a missile dedicated to the high-altitude area defense of vital targets, entered large-scale production and deployment, and surface-to-air missiles (SAMs) were deployed for the first time. These were followed by SA-3 Goa launchers. These weapons were assigned to the Zenith Missile Troops, the most important element of the air-defense forces, who were to defend Soviet command, industrial and

ABM-1B Galosh, possibly the only operational ABM in the world, pictured in travelling mode.

military targets against an enemy air attack. Fighter planes were assigned the lesser duty of intercepting attacking aircraft on the distant approaches to the defended areas.

Air Defense in the 1960s and 1970s
During the 1960s the Soviet defense system relied on the YAK-28 interceptor and the SA-5 Gammon missile, which provided a good high-altitude defense against enemy bomber attack, and on the MiG-25 Foxbat, which could operate at high speeds and high altitudes. The system had been designed to counter the threat posed by the American B-70 and the Skybolt missile, both canceled years earlier. While it may have been a useful instrument to counter the US SR-71 reconnaissance aircraft, the Soviet defense system was not adequate as a counter to the low-altitude-penetration tactics the United States had begun to emphasize. This recurring shortfall of the air-defense forces is symptomatic of the problems of the Soviet air defense system.

Qualitatively, the lagging effectiveness of the APVO and the ZMT is an excellent illustration of how shortcomings in the basic technological capacity of a country can limit the range of military options it has available, possibly at the expense of its military effectiveness. The Soviet lag in certain advanced data-processing and sensor technologies, which prevented it from fielding promptly the interceptors and airborne warning-and-control aircraft (AWACS) that are necessary to find and kill bombers at low altitudes, was a major factor in limiting Soviet effectiveness against the American bomber force through the 1970s. (The United States was able some years ago to deploy an AWACS aircraft and fighters whose radar is capable of discriminating and tracking aircraft flying near the surface without losing them in the ground clutter that would be picked up on a regular radar.)

The sheer size of the Soviet Union is also a severely limiting factor. Many thousands of radars, interceptors and surface-to-air missiles must be deployed for long periods of time to cover adequately the important areas in the USSR. Such a vast defense network required years to complete, and the United States, meanwhile, can devise new tactics or penetration aids to offset new Soviet developments. To aggravate all the other problems, the advent of nuclear weapons drastically reduced the number of penetrations that the defender can withstand. The air-defense forces have depended on a very restrictive ground-based control system that detects enemy bombers and directs the interceptors to their targets. Precursor strikes by either ballistic missiles or bomber-launched missiles could seriously degrade the Soviet air-defense network by destroying these ground-based control centers or interrupting their communications.

Although the Soviet air-defense system is not well prepared to counter the US bomber force, it probably has a formidable capability against the strategic bombers of such regional forces as those of Great Britain, France and China. And in the late 1970s the USSR was finally able to deploy new radars and low-altitude interceptors that would be effective against low-level-penetration attacks. The SA-10 missile, which was probably designed to intercept high-speed attack missiles carried by the US B-52 and FB-111 and the bombers themselves, is also now deployed.

Ballistic-Missile Defenses

Soviet interest in ballistic-missile defenses, which dates back to the 1950s, is an extension of the Soviet adherence to an active defense doctrine. Soviet ABM deployments preceded those of the United States and have been oriented toward defending administrative and economic centers rather than protecting land-based strategic forces, as the US program has been. Because they were originally designed to protect political centers and population and industrial areas, Soviet ABMs may have had different technical requirements than if they had been built for protecting ICBM sites. The most obvious difference would be the long range needed to intercept incoming missiles before they detonated near the area that was being protected. The major adversary for a Soviet ABM would probably have been submarine-launched missiles whose closer range would allow a steeper reentry angle and slower reentry speed than those of most ICBMs. Regardless of what their target was, the Soviets still had difficulty in developing an ABM system that would work.

Soviet ABM testing began in 1962 with the eight-ton Griffon. Although deployment of the system began around Leningrad, it was stopped the following year, and the system was dismantled by 1964, possibly because of the lack of adequate data processing and poor missile performance. It was undoubtedly deployed as a defense against the US Atlas and Titan missiles, as well as the Thor and Jupiter missiles based in Europe. The Galosh system began construction around Moscow in 1964 in locations similar to those of the first SA-1 SAM sites eight years earlier. Four complexes, each with target acquisition radars, tracking radars and 16 ABM launchers on soft sites, were installed by 1968. They would have been in response to the US Polaris A-1 and Minuteman I deployments in 1960 and 1962. The Galosh is a slow-reacting, 36-ton, exoatmospheric missile. The improved version can reportedly loiter in space by stopping and restarting its engines, thus giving its land-based radars time to discriminate between live warheads and chaff or decoys. This system would be an appropriate response to attacks by US Polaris A-3 and Minuteman II missiles equipped with multiple warheads and penetration aids.

The weak link in the Soviet ABM system has clearly been the radars. The missile-defense forces have relied on Hen House phased radars for the initial detection of a missile attack, and a combination of Dog and Cat House phased-array battle-management radars, along with the mechanically scanning Try Add engagement radars to perform the actual task of missile interception. The slowness of the Try Add radars and of the data-processing computers that support the Dog and Cat House radars leaves the entire ABM system inadequate against a high-intensity attack in a cluttered environment. All of them are vulnerable to the electromagnetic and blast effects of nuclear weapons. In the 1970s the Soviet Union began deploying large phased-array radars on the periphery of the country and expanded the use of early-warning satellites for earlier missile detection.

The 1972 ABM treaty, as amended in 1974, limits the Soviet Union to one ABM site with 100 launchers. The USSR chose to retain its 64 Galosh launchers around Moscow. Recently it removed 32 of them, perhaps in anticipation of a new ABM system in testing since the 1970s. The remaining launchers provide an adequate defense only against a small unsophisticated missile attack. The Soviet Union's motive in signing the ABM treaty has been a matter of debate, since the USSR initially showed great reluctance to consider limits on ABMs. There are a number of possible explanations for Soviet agreement to the treaty. The Soviet agreement not to deploy any new strategic anti-ballistic-missile defense systems was probably prompted by the persistent shortcomings of the Soviet strategic defensive force in general. Research and development on ABM systems have continued at a vigorous pace. Yet in none of the defensive programs was there sufficient progress during the 1970s to meet defense requirements. Technological limitations, American actions and the inherent difficulties of the problem continued to obstruct the strategic defense program.

In signing the ABM treaty, Soviet political leaders – if not military leaders – may have been accepting the Western concept of deterrence based on 'mutual assured destruction.' Or they may simply have been recognizing the inferiority of their ABM technology to that of the United States. But Soviet leaders may have concluded that they could best ensure the USSR's future ability to strike the US Minuteman force by preventing US deployment of an effective ballistic-missile defense system. The uncertainty of providing an effective ABM defense of its own urban centers, and the increasing capability of the Strategic Rocket Forces for performing its strike mission against US land-based strategic forces, may have led the USSR to conclude that the ABM treaty was a net advantage in terms of Soviet strategic doctrine.

5. ANTI-AIRCRAFT SYSTEMS

Modern SAMs have to be mobile to accompany a field army.
This US Army test firing was of a wingless rocket, unlike the
Chaparral system actually used.

In World War I even the anti-aircraft gun had scarcely matured as a weapon, and the seven known groups that tried to improve on the gun's average of 19,000 rounds per aircraft shot down were crippled by the fact that missile technology did not exist. In World War II in most theatres the average number of rounds fired per kill was even higher than 19,000, and in Nazi Germany the onslaught of Allied bombing was so relentless that the first SAM (surface-to-air missile) systems originated there in 1944. Some, like the German army Rheintochter (Rhinemaiden), had started even earlier, but it was 1944 before effective hardware began to appear.

By the end of the war the Germans had taken seven SAM systems a long way toward operational deployment, and Hs 117 Schmetterling (Butterfly) was in full production and deployment with LET 700, the Luftwaffe research and development flak unit. This rocket had radio-command guidance, a method by which a ground operator steers the missile via a radio link to coincide with a future position of the target. Schmetterling even had a proximity fuze, which detonated the warhead at its nearest approach to the target. Thus all the basic elements of an effective SAM were there in 1945 except for experience and reliability.

Britain developed a more advanced system, codenamed Brakemine, in which a winged rocket missile automatically flew up the center of a pencil-thin radar beam locked on the target, but lack of funding after the war's end caused this program to wither despite very encouraging tests. In the USA little was done until 1944 when the Japanese Kamikaze attacks suddenly magnified the need for better defense of ships. Two big projects, Little Joe and Little Lark, were rushed along at high priority, both with visual tracking and radio-command guidance. In 1945 Lark downed a radio-controlled F4F Wildcat target aircraft, but after VJ-Day it was thought better to start again. The British made the same decision about their very similar system, the Fairey Stooge.

Where the US Navy was concerned, the fresh start was a far-reaching research and development program, Project Bumblebee, managed by Johns Hopkins University. When industrial contractors were selected they became known as Section T, which led to names beginning with this letter for the missile systems that resulted: Terrier, Talos, Triton, Tartar and Typhon. Terrier was a rocket, while the bigger Talos was a

Top: The Terrier twin launcher of USS *Leahy* (CG-16) in January 1976.
Above: Twin launcher loaded with long-range Talos SAMs aboard USS *Chicago* (CG-11), in the Gulf of Tonkin in December 1967.
Right: The 88th XSAM-N-2 Lark ship-to-air missile of the USN, pictured in April 1949.

ramjet – the air-breathing, kerosene-fueled engine popularly called 'the flying stovepipe' – and both had four pivoted wings amidships which steered them up the center of a radar beam from the ship. Both went into wide service, and Talos survived to 1980 after being used in the Vietnam war against surface as well as aerial targets.

From Terrier and its shorter-range partner, Tartar, came a whole family of missiles known as the SM (Standard Missile) family. Even today these are the chief SAMs of the US Navy, and though even these newest versions date from 1970 they will continue in wide use into the next century as part of the Aegis system. Under RCA as prime contractor, the Aegis system has been created at great cost to give major warships comprehensive defense against sea-skimming missiles (see chapter 6) as well as aircraft. As installed in the now-building DDG-47 ships – which are called destroyers, though larger than World War II cruisers – Aegis comprises the completely new SPY-1A phased-array surveillance radar combined with several other radars and a vast computer system which detects targets, selects those posing the greatest threat, selects defensive missiles, fires them and guides them to kill the threat. SPY-1A has no moving scanner of the usual kind but instead 'sees' through flat panels, each about as large as a living-room wall, which face in all directions around the ship. The radar beams are steered electronically at the speed of light, and can guide a

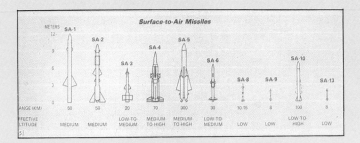

Far left above: SA-1 Guild on display in Red Square.
Far left below: SA-2 SAM being launched.
Facing page right: The number of Soviet SA-2 SAMs built exceeds that for all other surface-to-air missiles.
Left: SA-2 on its launcher.
Above: Chief types of Soviet SAMs.
Below; SA-3 Goa SAMs mounted on their mobile launcher in Egypt.

Standard Missile against an aerial threat as it flies faster than sound over a distance up to 19 miles for the MR (medium-range) version or 35 miles for the ER (extended-range). Of course, 35 miles is beyond the visual horizon to crew on deck.

The incredibly long lifetime of the Bumblebee/Terrier/Standard system, which now appears likely to outlast the typical human being, has deflected this brief history from the totally different US Army system devised to protect territory. This stemmed from the excellent, but large and costly, M9 radar fire-control system developed at the end of World War II by Western Electric's Bell Telephone Labs to control 90-mm AA guns. This locked one radar beam on a target aircraft and then, via a precise synchro link, aimed the gun(s) at a point where the target was predicted to be when the shells reached it. It was a good system – except to a field army on the move – but once a shell had been fired it could no longer be steered. The answer was to design a guided missile to fit a modified form of the same system, using the original contractors but with the missile handled mainly by Douglas Aircraft.

Designated SAM-A-7, and later named Nike Ajax, this became the first SAM to get into full combat service at a site near Washington DC in December 1953. The missile was long and thin, and with its tandem-boost motor (needed to blast it off the launcher) was almost 35 feet long, weighing 2455 lb. After the booster had burned out and dropped off, the Ajax missile continued on a liquid rocket burning nitric acid and aniline (today a solid motor would be considered preferable). As it climbed it was tracked by an MTR (missile-tracking radar) which, via a large and complex computer packed with fragile vacuum tubes, steered it to a future predicted target point. Compared with the wartime gun system, the big difference was that control was retained all the way to the target. If the latter were suddenly to change course, its new heading would still be followed by the target-tracking radar, which would keep changing the predicted future impact point. At the last moment, as the missile reached the target, its warhead was detonated by a sudden change in the complex pulse coding in the radar beam. Some Ajax missiles had three separate warheads, each resembling an aircraft bomb but surrounded by quarter-inch cubes of hard steel to form a lethal cloud of metal spreading outward at about twice the speed of a pistol bullet.

The worst feature of Nike Ajax was that it was not mobile, and each site involved thousands of tons of steel and concrete. The Army had about 16,000 Ajax missiles by 1958, in 40 battalions around the Free World. Many Allies used it, and a few installations survived to the late 1970s. So vast was the investment in Nike Ajax that the next generation had to fit into the same infrastructure, and the main thing about Nike Hercules was that all the speeds, heights and effective ranges were multiplied by about three. The Hercules missile is a monster, more than 40 feet long and weighing almost 11,000 lb. Under favorable conditions it can fly 87 miles, and its effective ceiling of 150,000 feet is about three times the height at which most aircraft can fly. New Hercules missiles were being made in Japan as recently as 1980, and there are many around the world which are being refurbished daily. The US Army, however, planned to replace it with a totally new system, at first called AADS-70 for Army Air Defense System 1970, and phased out its last Hercules troop in 1974, except for four batteries retained for training purposes in Florida and Alaska. AADS-70 did not come to be, as described later, but there was one further generation of the Nike system. This began as Nike Zeus and later became Nike X, culminating in the ABM Safeguard as already described.

Back in 1945 Boeing had been contracted to research a GAPA (ground-to-air pilotless aircraft) which brought in Roy Marquardt's new company for ramjet propulsion at supersonic speed. By 1949 this had led to the concept of an interceptor which could remain at instant readiness and, when needed, blast up vertically to cruise at several times the speed of sound for long distances and thus provide true area defense. In 1951 the USAF placed a full contract for what had become Bomarc, from Boeing and Michigan Aero Research Center (of the University of Michigan). After a giant development program, two generations of Bomarc defended large areas of the northern US, while other installations were bought by Canada, which perhaps unwisely scrapped its own piloted Arrow interceptor. All this matured in the 1957-59 era when it was a popular view that manned interceptors were obsolete. In fact, though Bomarc was pretty deadly out to ranges of 440 miles, it had a number of basic faults, quite apart from the fact that it is cheaper to use an airplane that can fly more than once.

No country believed more sincerely in the obsolescence of manned fighters than Britain; in fact, the RAF was told it would have no more manned aircraft of any sort in a famous government document of April 1957. British fighter airplanes were canceled, and the SAM systems were left in sole charge. The Royal Navy had a radar beam-rider resembling a bigger edition of Terrier. This system, Seaslug, had by 1961 demonstrated an SSKP (single-shot kill probability) of 92 percent, remarkable in that era of bulky electronics full of delicate vacuum tubes. In the 1960s Seaslug 2 came in, fitting the same ship magazines and launchers, but with much higher flight performance and better accuracy in adverse conditions such as effective enemy ECM (electronic countermeasures). A few Seaslug 2 systems are still in use, and in the Falklands campaign a Seaslug ship, HMS *Glamorgan*, was hit by a shore-launched Exocet missile against which the old SAM was not designed to be used.

Later the Royal Navy received a complete new-generation SAM, British Aerospace Sea Dart, which uses a ramjet missile resembling a smaller edition of the US Navy Talos. Air-breathing propulsion burns less fuel than a rocket, so the missile can stay under power to much greater distances. This has the advantage of better maneuverability, because when a rocket missile has burned out its fuel it slows rapidly, and if it then meets its target it may not be able to turn tightly enough to get within lethal range. Sea Dart can stay under power for at least 50 miles, which in any case is about the range

The massive ramjet-propelled SA-4 Ganef has long been thought to have a range of 47 miles, but a recent estimate is 150 miles!

limit for mast-mounted ship radars. Sea Dart is a standard long-range weapon on Royal Navy destroyers, and in the Falklands its presence denied Argentine aircraft the entire upper airspace for attack or reconnaissance and forced aircraft to come in at mast height. Sea Dart was not designed for point-blank low-level defense, but in fact when four Skyhawks attacked HMS *Coventry* she fired her twin Sea Dart launcher and downed two within seconds. But it takes time to reload large missiles when working to peacetime regulations (the rules were modified the same day), and the ship was promptly hit by the other two Skyhawks. Sea Dart scored eight confirmed kills, some under very tough conditions, and has also been developed in a lightweight containerized form for such small ships as fast patrol boats.

In the Soviet Union the first generations of SAMs were predictable radar-guided rockets, the main point being that the mass-produced SA-2 Guideline, counterpart to Nike Ajax, was put on wheels from the start. Soviet armies have always thought less in terms of steel and concrete than in terms of how to surge across mountains and rivers – mobility. Admittedly, the complete SA-2 system, with massive rotating launchers, Side Net heightfinders, Fan Song target-track radars and many other items, was once calculated to weigh roughly 100 tons, but it can still be moved from place to place, and without any paved roads. In the 1960s there were over 4000 SA-2 launchers in Soviet client states all over the world, and hundreds of these massive missiles were fired in Vietnam.

SA-2 was designed in 1949-52, and American assessment often overlooked subsequent developments. American aircrew had the feeling that a SAM was a thing to dodge, certainly something to spoof with electronic countermeasures, render ineffective by jamming and hit with such ARMs (anti-radar missiles) as Shrike. This was a fool's paradise, because Soviet designers did not go on holiday in 1952. In 1964 a parade through Red Square included monster ramjet SAMs, which NATO promptly called SA-4 Ganef, and though these have not yet been encountered, there is no reason to doubt their lethality in providing large-scale area defense to field armies. They closely resemble a Sea Dart carried in pairs on an amphibious armored chassis which can go anywhere the armies can go, and the protection they give is about 100 percent out to 47 miles. This is something the US Army has never had.

The next Soviet SAM, SA-5 Gammon, has been covered under strategic systems. So we come to the new tactical SAM which rumbled across Red Square on 7 November 1967 in triple batteries on go-anywhere amphibious carriers. NATO called it SA-6 Gainful, but perhaps did not examine it too closely, beyond noting that the missile has an integral ramjet engine and thus flies faster than sound all the way to the target, which can be as much as 37 miles away. The triple launchers and their reload vehicles are accompanied by neat but powerful Straight Flush radars, as well as big Long Track surveillance radars to give early warning of hostile aircraft. It seems that nobody in the Pentagon, at least, had ever considered that some investigation and counteraction might be necessary.

Then in October 1973 the Yom Kippur war broke out, as Egypt thrust deeply toward Israel. At once the highly skilled Heyl Ha'Avir, the Israeli Air Force, thundered into action, but its modern aircraft – Mirages, Phantoms and Skyhawks among them – fell in droves. Here was a SAM which did its job; it was impossible to outmaneuver, and the whole panoply of Western countermeasure systems had no effect on it. This was really inexcusable, because the clue to SA-6 lethality was use of a CW (continuous-wave) radar, which had been common in NATO countries for years and, in fact, is a feature of the USA's mass-produced Sparrow AAM (air-to-air missile). But nobody had produced any ECM to defeat CW signals, so the Israelis suffered very severe losses, though they filled the sky with chaff and did their best to avoid the highly mobile SA-6 troops.

Today SA-6 itself has had to be updated to try to keep abreast of the West's anti-CW countermeasures. This is not the situation with the next Soviet SAM, SA-7 Grail. The idea of a SAM that could be fired by a foot-soldier was studied in 1948, but the USA/USMC feasibility awards were dated 1958 and initial operational service with the resulting missile, Redeye, was not until 1968. The Soviet SA-7 was similar in timing and technology. Both use a shoulder-mounted tube with a sight (and, one hopes, an IFF – identification friend or foe – to

Left above: A Soviet artilleryman aims his SA-7 Grail SAM.
Left; Deadly SA-6 Gainful ramjet-driven SAMs on exercise.
Above: SA-9 Gaskin SAMs on the march, with ZSU-23-4 flak
vehicle in the rear.
Below: The SA-8 Gecko vehicle is air-portable and amphibious.

avoid shooting down too many friendly aircraft). In the tube is a slim missile with an IR (infra-red) seeker in the nose which can detect and lock-on to any suitable source of heat. Unfortunately, in these early man-portable SAMs the only suitable source was the jetpipe of the attacking aircraft, and this could not be seen clearly until the enemy aircraft had made its attack and was receding at a rate of knots. Only then could the operator get a green light in his sight system telling him the seeker had locked on and that the SAM could be fired by pulling a trigger. Even then the small rocket might have severe difficulty in overtaking the target within its range limit of about two miles.

For these reasons Redeye was given only a qualified welcome by American troops, while the contractor, General Dynamics, Pomona (California) Division, went on to invent a much better SAM called Stinger, which can not only be fired at an oncoming aircraft but can also home onto sources of two kinds of radiation, IR or UV (ultraviolet). As for the Soviet SA-7, this at least had much greater range, because a jet near Oman was hit at a slant range of over six miles. A British-made

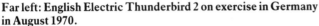

Far left: English Electric Thunderbird 2 on exercise in Germany in August 1970.
Left above: The RAF will continue to use the Bloodhound 2 air-defense missile through the 1980s.
Left below: Thunderbird test firing at the Aberporth range, Dyfedd (Wales).
Above: Switzerland is another Bloodhound 2 operator.
Right: Hawk missiles of the US Army are fired from a triple launcher.

Hunter near the Yemen was hit at a height of 12,000 feet above the ground, yet the SA-7 missile is only fractionally heavier than Redeye at 20.3 lb. SA-7 has caused ripples of disquiet because the number of launchers (presumably all loaded) is conservatively put at 120,000. Of these at least 10 percent have, by accident or design, gotten into the hands of irregular forces, including many dissident or terrorist groups. The SA-7 operator waiting under the approach path of the 747s at a major airport is a popular scenario among TV writers, but less popular with the airport authorities.

This problem of clandestine weapons is bound to get worse, because a key factor of modern technology is that the hardware needed to do most things keeps getting smaller. Today the success of the SAM has been such that in any future war the entire sky will be empty except that part of it below the tops of tall trees. Flying 'under the radar' seems to be the only way to survive, and this automatically downgrades the importance of big SAM systems designed to kill at a distance. When Britain built its first major SAMs, it assumed the targets would be jet bombers at high altitude. The British Army received two generations of Thunderbirds, filled with heavy engineering and massive vehicles, while the RAF deployed Bloodhound, almost in the class of Bomarc. In fact, Bloodhound is an excellent weapon, with twin ramjet engines giving a range of 'several score miles' and homing automatically onto the radar reflections from the target (a method called SARH, semi-active radar homing). Bloodhound is so effective it has

never been junked, and many are still at readiness in England, Switzerland, Sweden and Singapore!

But since the mid-1950s it has been self-evident that the future need would be to kill the low-level attacker. By far the most widely used, as well as the oldest, of these systems is the US Army's Hawk, a good name coined from Homing-All-the-Way Killer. Raytheon was prime contractor for this radar-guided SAM, which uses a 17-foot missile weighing 1380 lb – far bigger than the systems developed subsequently. Hawk went into service in 1959 and has since been progressively updated; despite its size, over 39,000 rounds have been delivered. Nominally portable, the Hawk system invariably operates from big fixed sites which could not be moved readily.

Surprisingly, the new US Army system developed to replace Hawk, as well as Nike Hercules, is in many respects just as big as its predecessors and many times more costly. Originally called AADS-70, it still had not reached the Army by 1980, by which time it was called Patriot. Key to this SAM system is a very large and extremely advanced radar of the phased-array type, in principle like those on the DDG-47 warships in the Aegis system, which replaces nine separate radars in the US Army's existing SAMs. This incredibly complex computer-controlled radar scans the entire airspace and handles many targets and many missiles simultaneously, a unique feature being TVM (track via missile) which combines a downlink from the Patriot missile to the ground, even though the missile has its own seeker head and homes automatically on the target. In the early 1980s the Patriot system was slowly taking an effective place in the US Army, but cost will inevitably reduce the numbers.

Delays on Patriot resulted in the mass-produced Sidewinder AAM (described later) being adapted as a mobile US-Army SAM called Chaparral, quite effective at very close ranges. But Chaparral was intended only as a stop-gap, and in the early 1970s the Army cast about for a replacement and after much testing picked the

Below: Test firing of the much-delayed Patriot SAM, long planned as the next-generation SAM of the US Army.
Right: Test firing of a US Roland when this was still a full-scale program for the US Army in 1979.
Right below: Rapier launcher of the Canadian 12th Air Defense Regt on Exercise Reforger in 1977.

Franco-German Roland in 1975. In France and Germany this 139-lb missile, fired from a two-barrel turret on an armored vehicle, has been made to work reasonably well. The only British aircraft shot down by a SAM over the Falklands was a Harrier downed by a Roland vehicle bought by Argentina (later, Euromissile, makers of Roland, claimed this missile had shot down four Harriers and damaged a fifth, but the British Ministry of Defence showed that in fact only a single Roland missile had been fired). But in the US Army Roland proved a major error, with time and money wasted on a prodigious scale until it was decided to cut the losses and field just 10 Roland systems with the Rapid Deployment Force, instead of the planned 14,000 rounds.

The direct rival to Roland, and much more successful, is the British Aerospace Rapier. Developed in the 1960s and first delivered to the British Army and RAF Regiment (which handles RAF airfield defense) in 1967, Rapier is designed to a new concept known as a 'hittile.' This relies upon a guidance system so accurate that it can be guaranteed that the missile will actually hit the aircraft. Thus the warhead, exploded inside the target, can be much smaller and triggered by a simple contact fuze instead of a proximity fuze. In the case of Rapier, this has enabled the missile to be made so small that it weighs only 94 lb and thus can be man-handled. Early versions were fired from a quad launcher and steered by a TV optical link that is so accurate the missile has consistently passed dead-centre through Rushton targets with a diameter of just 7½ inches. In bad weather it is locked to the target by computer-

generated Blindfire radar, with the same commands transmitted over a microwave link. The entire system is miniaturized and air-portable. In the battles over the Falklands, every Rapier, launcher and tracker arrived after an 8000-mile voyage as deck cargo which often included many severe impacts and mistreatments much beyond the design requirements. Then the weapons were bundled ashore in landing craft or slung under helicopters and rushed straight into action to fire at high-speed jets, often at a level below that of hill-mounted Rapiers, with friendly troops and helicopters in all directions at close range. Often the attackers were masked until the last moment by hills and low cloud and visible for only three to seven seconds. Under these conditions 14 confirmed kills were scored, some of them in newly tried techniques in which the missile was fired and then turned in mid-flight to engage the attackers.

One of the many customers for this deadly missile is the US Air Force, which uses it to protect its airfields in the UK and is likely to use it in Germany also. The latest version is Tracked Rapier which consists of an eight-shot launcher mounted on an M548 armored amphibious launcher. The vehicle can be driven out of a C-130 in a ready-to-fire configuration, and can fire within 30 seconds of coming to a halt when traveling cross-country. Tracked Rapier is judged to give the British Army the world's best protection against low-level air attack.

Another British Army SAM is Blowpipe, an infantry weapon which was used by both sides in the Falklands

Left; US Roland will be used to cue Redeye (left) and Chaparral (right) SAMs.
Left below: Blowpipe SAM of the Royal Marines at San Carlos Bay, Falklands, in May 1982.
Above: Short's Blowpipe SAM being fired.
Below: General Dynamics Pomona has developed Stinger as a far superior successor to Redeye.

war, scoring eight confirmed kills. This automatically follows the line of sight of the optical unit on the launcher. There are many similar man-portable SAMs, including the French SATCP and Sweden's RBS 70, which flies up a laser beam. Such weapons make the design of costly battlefield helicopters with a missile sight low on the nose (necessitating the exposure of the whole aircraft to steer the missile) appear extremely shortsighted.

Returning to naval SAMs, Britain struck a blow for low-cost simplicity with Seacat, developed in 1958-62 as a radio-command missile for close-range defense. Later it got an add-on radar for use at night or in bad weather, and a land version was called Tigercat. All sold widely to many countries, and in the Falklands the Argentine army used Tigercat while RN Seacats downed six attacking aircraft. But a far more deadly weapon was developed from 1969 specifically to kill sea-skimming missiles as well as supersonic aircraft. At first called GWS.25, it became known as Seawolf. Each missile is treated as a round of ammunition, and in the basic system in service since 1976 a six-barrel launcher is loaded by hand. When a threat enters the area of a Seawolf ship, a computer evaluates the threat, assigns a missile and slews the radar and launcher. Firing is automatic, but if the radar has difficulty tracking a target at extremely low level, the task can be taken over

manually by a boresighted TV camera. Typically the target is destroyed 10 seconds after the threat is detected. Seawolf has demonstrated 100 percent lethality against every kind of target, and has successfully intercepted and destroyed 4.5-inch high-velocity naval shells. In the Falklands Seawolf ships numbered only two, and both were carefully avoided, but despite this Seawolf systems did get a chance to fire on five occasions, with five kills photographically recorded. The next version uses the same missile in a battery of 30 vertical launch tubes, all instantly available, to cope with saturation mass attacks. Another version is packaged into commercial containers. Pre-load containers can be stored until a time of emergency and then loaded inside an hour onto a merchant ship, converting it into a SAM platform with formidable firepower.

So far all the anti-aircraft missiles covered have been SAMs. There is also the large category of AAMs, already briefly referred to in the case of Chaparral. An AAM, by definition, flies from one aircraft to another, and this in theory ought to make its problems simpler. It is probably already moving through the air at high speed when it is fired, instead of having to start from rest, and the target is clearly defined and isolated. Such a target is ideal for radar, and as the target is probably a jet it is also a very powerful emitter of heat (IR) radiation, toward which it is relatively simple to make a missile fly. Nevertheless, the first AAMs in World War II were German weapons steered to their targets by signals passed along fine wires unwound from bobbins on the wingtips. Both X-4 and Hs 298 almost got into action, Hs 298 also having a radio-guided version, but they were too late to stave off defeat.

Development was picked up at once in the USA, where various forms of guidance were tried in such missiles as Meteor, Firebird, Gorgon and Oriole. Much progress was made with SARH, in which the fighter radar 'illuminates' the target and the missile homes on the reflected radiation, as well as on active radar homing in which the missile carries its own radar in the nose –

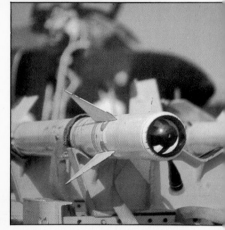

which means a big missile. In fact, the first missiles to go into production were both very small and a triumph of ingenious design and dense packaging.

The first was developed by the young Hughes Aircraft to form part of the same company's E-9 radar fire-control for the USAF F-89H Scorpion and MG-10 for the F-102 Delta Dagger. Called Project Dragonfly, then XF-98, then GAR-1, it eventually became AIM-4, with the name Falcon. The first SARH version was cleared for production at a new plant outside Tucson in 1954, and the first IR Falcon followed in 1956. These weighed only a little over 100 lb and were about 6 feet 6 inches tall but only 6½ inches in diameter. By the late 1970s about 60,000 Falcons of many versions had been made, but few remain.

The other pioneer small missile was Sidewinder, and it is probably the most numerous and most copied missile in history. Certainly if one includes the Russian, Chinese, Israeli, South African and other copies, the total must approach half a million. More significantly, it got AAM designers away from trying to build impressive giant missiles and started them thinking more in terms of bits of stovepipe. The concept was due to a small team at NOTS, the US Naval Ordnance Test Station at China Lake, California. In 1949 it studied the problems of building an IR seeker head. This calls for a sensitive cell, which in those days was made of lead sulfide PbS, which when refrigerated to a temperature much colder than that of a domestic freezer can become intensely sensitive to IR radiation. The team then had to package this tiny cell inside an optical system that

was basically a telescope to focus the weak radiation. This all had to go inside the head of the missile, behind a glass nose (as IR has a rather longer wavelength than visible light, the best glass may not look transparent). To make life much harder, the NOTS team elected to make their missile just 5 inches in diameter. This was in the days when electronics consisted of discrete components, vacuum tubes and soldered joints being made by hand.

Amazingly, the seeker head was made to work, and at the back of it were mounted four delta (triangular) fins power-driven to steer the missile. Then came about eight feet of gaspipe-like body containing the warhead and solid rocket motor, with four large tailfins bringing up the rear. Philco was brought in in 1951 as main industrial contractor and quickly delivered 80,900 of the first version, which was later designated AIM-9B. Today the same team at Newport Beach, California, renamed Ford Aerospace, is a leader in the Sidewinder team and is still mass-producing much better homing heads for today's versions. Early Sidewinders were reasonably satisfactory if the interceptor pilot could get close behind his target, in the traditional way, and aim the missile up the enemy jetpipe (a satisfactorily hot target). Occasionally, Sidewinders managed to do this but then failed to explode. In 1958 one slammed into a Communist Chinese MiG-15 near Quemoy Island and was brought back intact. In 1959 the first Soviet copy, called AA-2 Atoll by NATO, was seen under the wings of a MiG with red stars.

There were other problems that afflicted many other pioneer AAMs. Seeking the best source of IR, the missile might detect the sun and try to home on it, or onto its reflection in a lake or friendly greenhouse. It has taken almost 30 years to devise today's hi-fi seekers, some of which are of the so-called staring focal-plane-array type in which modified optics focus the radiation onto an array of super-sensitive cells, which give virtually 100 percent lock-on to the correct target and to no other. It is 20 years since press releases proclaimed that a missile could 'detect a cigarette at five miles,' but that

Left top: Falcon AAMs being fired from a USAF Convair F-102A Delta Dagger in March 1957.
Far left: The chief members of the Hughes Falcon AAM family.
Left: An AIM-9B, the original version of Sidewinder (with F-4 in the rear).
Left below: An AIM-9J Sidewinder on the wingtip of an Aggressor F-5E Tiger II.
Below: A Libyan Arab Air Force Su-22 Fitter armed with AA-2 AAMs.

was of doubtful value if it tried to lock-on to the wrong cigarette (or, literally, to anybody's cigarette).

We do not have full details of the latest AA-2-2 Advanced Atoll, which in any case has been overtaken by the much shorter and more maneuverable missile NATO calls AA-8 Aphid, carried by, for example, late versions of MiG-21 and MiG-23. But the original Sidewinder has been developed through ten major versions, each offering a new capability. One, AIM-9C, had SARH radar guidance by the Motorola Company for the Navy's F-8 Crusader fighter. All the rest had improved IR homing, and the latest AIM-9L and forthcoming 9M have visibly different long-span pointed control fins for better dogfight maneuverability. All recent models have had seeker heads that can lock-on over a wide range of angles, not just dead astern the target, and L and M are so-called all-aspect missiles which can lock-on to the target from any direction, including head-on. But the chief advance has been to improve discrimination of the IR seeker by cutting out response to the target's background and making it as far as possible proof against enemy countermeasures (IRCM, the heat-wavelength counterpart to ECM) such as intensely hot flares ejected by the target to throw the AAM off the scent.

The story of Sidewinder, which is by no means entirely over, mirrors the story of IR-homing dogfight missiles in all countries. Apart from the wire-guided anti-tank missile, this is the class of guided missile most nations have felt they could make for themselves. Most have the same five-inch diameter as the US weapon; amazingly, the first Sidewinder copy by the Chinese People's Republic made life even harder at 4¾ inches. Today a few have more room inside, including Israel's battle-proven Shafrir at 6¼ inches and France's battle-proven Matra Magic at 6½ inches. The Soviet AA-8 Aphid is truly remarkable, at just 4¾ inches combined with very short length, and drives home the lesson that Soviet microelectronics have reached an impressive level with production hardware. AA-8 was first seen in 1975 and many thousands are in use.

IR homing is by far the preferred method for close-range dogfight missiles, but it is no good for stand-off engagements. The heat radiation from any source falls away in intensity as the square of the distance, so three times the target range means one-ninth as strong a signal. On top of this, the radiation is attenuated by moisture and dust in the atmosphere, and while this is not a major factor at high altitudes it becomes very important at the low levels at which almost all modern air combat takes place. In a tropical rainstorm it is doubtful if an IR-homing missile could lock-on at a distance of a quarter-mile. Some other guidance has to be used, and the almost universal choice for many years has been SARH.

This was pioneered in the United States, but not many Western fighters can carry radar-guided AAMs. This is in the sharpest contrast to the interceptors of the Soviet PVO (Air Defence Force) whose AAMs all come in two versions, one with IR and the other with SARH. Every aircraft has at least two pylons, and it is normal practice to load AAMs in pairs with one of each kind.

Thus the pilot can always choose one best adapted to the conditions, or for maximum lethality he can fire both in quick succession.

Apart from Falcon, already mentioned, the first SARH missile in production was the US Navy's Sparrow. This originated at the Sperry company after World War II as a beam-rider. Sparrow II was canceled and the first important version was Sparrow III, for which the prime contractor has always been Raytheon. This is a substantial missile with a body diameter of 8 inches, a length of 12 feet and a launch weight of 450-500 lb. Its first major application was as the primary armament of the F-4 Phantom II of the Navy and Marine Corps, which carries four recessed into the underside of the fuselage. The Westinghouse radar, originally the APQ-72, is locked on the target at a range of perhaps 30 miles, and a missile can be fired at about 25 miles, homing on the radar signals reflected from the target. SARH has a big advantage in that the homing signal gets stronger as the target is approached, as it does with IR but not with beam riding. Sparrow maneuvers with four moving delta wings amidships, and at the nearest point the 66 lb warhead is detonated by a proximity fuze, showering

Left above: A Soviet MiG-23 swing-wing interceptor armed with AA-7 Apex and AA-8 Aphid AAMs.
Left: Firing an AIM-7 Sparrow medium-range AAM from an F-4B.
Above: A recent MiG-21bis armed with two AA-2-2 Advanced Atolls and two AA-8 Aphids.

the target with lethal stainless-steel fragments. The most numerous Sparrow III was AIM-7E, some of which were used in Vietnam (though the political rules of engagement calling for visual identification of the target made long-range interception impossible). Raytheon then rearranged the parts as a result of the introduction of compact solid-state guidance. The resulting AIM-7F has not only a much more powerful 88 lb warhead but room for a new motor giving almost double the range, to a limit of over 40 miles. This new Sparrow is also being made by General Dynamics at Pomona.

Sparrows arm the F-14, F-15, F/A-18 and Italian F-104S. There is no reason why they should not arm the F-16 Fighting Falcon, and Sparrows have been fired (but not yet guided) from that aircraft. But lack of

Sparrow capability has been suggested as the only visible reason for the last three major fighter sales battles (Canada, Australia and Spain) having all been won by the F/A-18.

Uniquely, the Navy's F-14 Tomcat also carries a third type of AAM even more powerful than Sparrow. Back in the 1957 era the Navy studied the possibility of using an AAM with range well over 100 miles to defend the fleet. Bendix created the great Eagle missile, with 127-mile range, but this was canceled along with its F6D Missileer carrier aircraft in 1960. The technology trickled on in the powerful Hughes AWG-9 radar planned for the canceled F-111B and finally bore fruit in 1964 when a completely new AAM, Hughes AIM-54 Phoenix, was given the go-ahead, flying the following year. In the F-14 the radar can simultaneously track 24 targets and engage any six, choosing if necessary six particular aircraft from a close formation at well over 100 miles' range. Positive identification is greatly assisted in newly modified F-14s by a Northrop TCS (TV Camera Set), giving a greatly magnified image of the target in all weather. Six Phoenix can be carried, each weighing 1008 lb and costing over $500,000. Each missile can fly to its own designated target, finally switching on its own small planar-array radar in the nose to home on the target unerringly. No AAM has ever equaled the long range of Phoenix, and new versions are now in production with cost-reducing features in the airframe and miniaturized solid-state electronics.

One AAM that might have been expected to beat Phoenix is the Soviet weapon known to NATO as AA-6 Acrid. Several Soviet AAMs have been large, but Acrid is enormous, over 20 feet long and weighing an estimated 1650 lb. Like other Soviet AAMs it comes in both SARH and IR versions, two of each being the normal armament of the massive 2000 mph MiG-25 Foxbat. It may be that Western analysts are (again) indulging in wishful thinking, but their estimates of AA-6 range are modest, 23 miles being a common figure. Why the range is so short has not been explained, and seeing an AA-6 interception over this distance on Allied surveillance radar or satellite does not prove this is the limit attainable. Using any reasonable internal packaging and rocket motor performance, a range well in excess of 100 miles is likely. At least it is known that the AA-6's estimated speed of Mach 2.2, as believed for years, is about 70 percent too low.

Most important of the medium-range Soviet AAMs is a fairly new weapon known to NATO as AA-7 Apex. It is in the class of Sparrow but is larger all round, and its extremely large delta wings would appear to give it better maneuverability at most altitudes. It is calculated to have about 30 miles' range in its SARH version, the IR model naturally being limited to about 10 miles. Reports of various other Soviet AAMs are hazy and of little value.

Britain began with an odd beam-riding AAM, Fireflash, whose two rocket motors were carried outside the body of the missile so that they could be discarded at burnout. Not many were built, but the IR-homing Firestreak sustained a large production program and is still found in small numbers on the Lightning intercep-

Right: Standard future interceptor of the RAF is the Tornado F.2 armed with Sky Flash and AIM-9L Sidewinder AAMs.
Below: Test launch of the advanced AIM-54C version of Phoenix from an F-14, at Point Mugu in 1982.

Above: Tigercat SAMs on a triple launcher; this missile was derived from the ship-based Seacat.
Right above: First firing of the AIM-120A Amraam from an F-16A.
Right below: The Matra R 530 was operational on the Mirage IIIC by the mid-1960s.

tor. The more common Lightning missile is Firestreak's next-generation successor, Red Top, which has the good range for an IR weapon of 7½ miles. To arm its Phantoms and the new Tornado F.2 interceptor, Britain took a license for Sparrow and designed a much more effective radar-seeker head of the monopulse type, and a new active-radar warhead fuze. The resulting missile, Sky Flash, went through its firing trials at Point Mugu, California, from November 1975. Each shot was calculated to try to defeat the missile guidance by providing the most difficult conditions possible, but the overall program was judged 'the most successful ever flown at Point Mugu.' Over half the rounds fired actually struck the target, while the miss-distance of the others averaged 'about one-tenth that of most radar-guided AAMs.' Sweden noticed this unrivaled performance and has placed a major production order followed by a repeat order.

In the USA the goal of a common USAF/USN missile is finally being realized with Amraam, Advanced medium-range AAM. After a tough flyoff competition, the prime contract was placed with Hughes, which has made rapid progress. Amraam is to be a launch-and-leave missile. One trouble with such SARH weapons as Sparrow is that to keep illuminating the target, the fighter has to keep flying toward it. Sparrow may have been fired at a range of 25 miles, but the range closes very rapidly. Time and again in simulated combat the 'enemy' has noticed the oncoming F-14 or F-15 at the

last moment and zapped off a Sidewinder just before the arrival of the Sparrow. This makes the score 1-1 instead of 1-0. The necessity for flying always toward the enemy is a major snag, and in a launch-and-leave AAM it is no longer a factor. Amraam is fired at ranges 'not less than those for AIM-7F Sparrow,' whereupon the fighter can at once turn tail and head for home. Amraam flies on inertial midcourse guidance until approaching the target, when it switches on its own miniaturized active radar and homes on the hostile aircraft in the same way as the big Phoenix. Unlike Phoenix, Amraam is barely half as heavy as Sparrow, and Hughes claim it will outperform Sparrow and cost less to build. Its resistance to ECM will be as good as that of the British Sky Flash. Its designation is AIM-120A.

When Amraam finally reaches the squadrons aboard all current US fighters, as well as the RAF Tornado F.2 (replacing Sky Flash in due course), it will offer a major capability not yet discussed. Today all aircraft penetrat-

ing defended airspace do so at the lowest possible level above the ground. This means that a defending fighter has to look down on them from above, and he sees the target very close to the intense radar reflection caused by the ground. Until recently no AAM could engage a low-level target. So-called 'look-down shoot-down capability' was urgently sought, and in Amraam it is fully obtained. Sky Flash is pretty good in the snap-down role, and like Amraam it can also be fired in snap-up maneuvers to intercept targets perhaps 30,000 feet higher than the defending interceptor at the moment of firing. The only other AAM that claims to have snap-up and snap-down capability is the French Matra Super 530, an SARH weapon carried by the Mirage F1 and 2000. It is a vast improvement over the same company's R.530.

While NATO nations have agreed that the US shall develop Amraam, the corresponding Asraam (Advanced short-range AAM) is being worked on by British Aerospace and the German BGT company, forming a team which was joined by Hughes in September 1982. British Aerospace has immense experience in this field, having spent over 10 years developing first Tail Dog and then its own Sraam, both of which had unrivaled dogfight maneuverability gained by using thrust vectoring of the missile motor. Details of Asraam are classified, and some have yet to be agreed upon, but the missile is designed to be wingless, to have an advanced IR seeker and to maneuver in seemingly impossible ways by motor vectoring. As Asraam is intended as the replacement for Sidewinder, its importance can hardly be overstated.

6. LAND TACTICAL SYSTEMS

A TM-61A Matador of USAF Tactical Air Command in November 1955.

In several respects missiles for use in land battles pose some of the greatest challenges. They have to be simple, tough and able to withstand the harshest treatment in all weathers. They must be not only mobile, but able to travel across country wherever an army can go, and the Soviets have a general rule that every army weapon system has to cross mud, snow, ice and soft sand and either wade deep rivers or be fully amphibious. Not least, tactical missiles must be able to hit small moving targets which are difficult to distinguish against the background – in sharp contrast to ships or airplanes – despite the presence of heavy and sophisticated hostile countermeasures.

There is some difficulty in knowing even what to include under this heading. Unguided artillery rockets and shells from guns are clearly out, but today plain rockets and artillery are becoming more sophisticated; when their projectiles reach the enemy, they seek him out, lock-on and home in, even if the original aim was faulty.

While there is no single narrative thread linking the diverse types of tactical missile, it will be noted that virtually everything in this category has been, and is being, done by the United States and the Soviet Union. Though there are long-standing rumors about missiles from Egypt, Israel, Taiwan, China and elsewhere, the only modern weapon in this class by a third-party is France's Pluton.

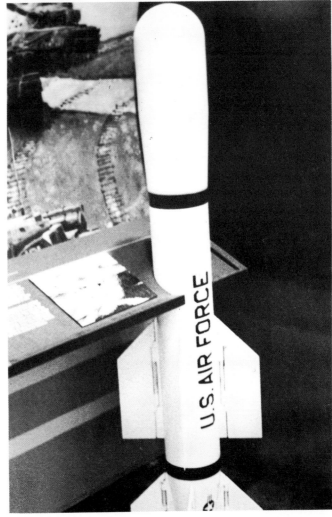

Left above: Honest John artillery rocket on its mobile launcher.
Left: Launch of a Lacrosse artillery missile at Ft Sill, Oklahoma.
Above: Members of the Bullpup ASM family, clockwise from top right: AGM-12, ATM-12A (training), AGM-12B (heavy) and AGM-12D (nuclear).
Right: Wasp is the latest US air-ground anti-armor missile, with millimetre-wave radar guidance.

United States

There is a massive and well-documented history of tactical missiles developed in the US from 1916 on, and several systems were mass-produced and were about to go into action at the time of the World War I Armistice. A little was done by the Navy in the late 1930s, but the story picks up again in 1944 with literally dozens of weapon systems ranging from explosive-packed B-17s and B-24s guided by radio, through copies of the German V-1, to purpose-designed missiles of advanced form.

In the 1950s the most important were the US Army's Corporal ballistic missiles, which lobbed a 30-inch diameter warhead with a nuclear or large conventional charge over ranges up to 86 miles, and the Air Force's TM-61 Matador and next-generation TM-76 Mace, which were really pilotless jet bombers that in their final form (1960) had mobile launchers stored in hardened shelters. Small numbers of the Army's massive Redstone ballistic missile, a distant descendent of V-2 with a range of over 200 miles, were deployed, while the Honest John artillery rocket was made in large numbers and many are still in use, though by modern standards they are obsolete.

Originally sponsored by the Marine Corps, SSM-A-12 (later called MGM-18A) Lacrosse was a 2300 lb rocket with complex radio-command guidance which could be handed on from one ground station to another, or to an observer in a helicopter. It carried conventional, nuclear or chemical warheads over ranges to 19 miles and was used by the US Army as well as the Canadian Army. A shortcoming was inability to resist enemy jamming and other countermeasures, and Lacrosse had a brief active life in the early 1960s. In contrast, MGM-29A Sergeant, the successor to Corporal, had a long service career and was still in front-line duty in 1978. It was a massive and complex system, but still a big improvement over Corporal, with at least some cross-country capacity and the ability to be fired within an hour of selecting a site. The missile was fired at 75 degrees and used self-contained inertial guidance over ranges up to 87 miles.

As the important Pershing I and II missiles and the GLCM version of the Tomahawk cruise missile have been covered in the chapters on strategic systems the only other relevant US weapon actually in service is Lance. This cost/effective program fills what would otherwise be a very serious gap in Western defense, because it alone provides the mobile firepower over long theater distances vital to success in any future major land battle. The original basic system is contained in two air-portable amphibious tracked vehicles of the M113 family, while Hawker Siddeley Canada has produced a lightweight launcher which makes the system heli-portable and air-droppable by parachute.

The Lance missile, MGM-52, is just over 20 feet long and weighs 3373 lb with the usual 10-kiloton M234 warhead. In its basic role as a corps-support weapon, the Vought-developed Lance replaced both Honest John and Sergeant, with several quantum jumps in flexibility, accuracy and effectiveness. Rocketdyne provides storable-liquid propulsion, and guidance is inertial. Launch pictures show dense smoke from the four ports of the spin motor, which is actually the exhaust from the on-board gas generator. Most of the flight is a zero-g ballistic trajectory, and the sustainer thrust is smoothly varied by throttling to achieve the exact cutoff velocity for the target range. Alternative warheads include a conventional type, a chemical type, a DME type with precision Distance-Measuring Equipment, and a large 1000-lb cluster dispenser. An ER (Enhanced-Radiation) nuclear warhead for use against massed armor has been developed, and the newest warheads use the main missile as a 'bus' from which are dispensed TGSMs (Terminally Guided Sub-Munitions) with individual IR-seeking guidance, as briefly noted in the preceding chapter.

Today a major increase in tactical firepower for the US and probably other NATO armies is expected from MLRS (Multiple-Launcher Rocket System). Though basically unguided, MLRS is so advanced in concept as to merit discussion here; to the basic 'bus' missile there will in future be a TGW (Terminally Guided Warhead) giving precision attacks not only against armor, but against all battlefield targets. MLRS, by Vought Corporation, rides on a self-propelled launcher-loader

Left: The Vought T-22 Corps-support 'bus' missile guided submunition may be procured for the US and allied armies.
Top: Vought Lance being fired from its tracked amphibious launcher.
Right: Complete system mock-up of the MLRS rocket system.

which one man can resupply with 'six-pack' boxes of fresh rockets of 227-mm (8.94-inch) caliber. The warhead originally fitted is Phase I, containing 644 antipersonnel and anti-armor bomblets. Phase II is a German mine-dispensing system called AT-2, using a rocket increased in caliber to 240 mm (9.45 inches). Phase III uses the TGW, and in February 1983 it was announced that this will be developed by a powerful multi-national team under GD Pomona (USA) and including British Aerospace Dynamics, Sperry, Dynamit Nobel, SEP and Scicon. TGW may use both millimetric radar and an IR sensor for precision homing onto targets over the final 1¼ miles of a 25-mile flight.

Many other countries use high-firepower artillery rocket systems. None is known to come anywhere near the capability of MLRS in its three planned phases.

France
Developed for the French army by a large industrial team led by the Tactical Missiles division of Aérospatiale, Pluton is a nuclear missile for use over ranges up to 75 miles. The basic projectile is 25 feet long and has a diameter of 25.6 inches, the launch weight being 5342 lb. The standard warhead incorporates the same 25-kiloton charge as the AN.52 bomb used by the French AF (Armée de l'Air). In its present service form it is deployed singly on AMX.30 tank chassis, each round being fired straight out of its box container. Guidance is by a simple strapdown inertial system, the CEP (Circular Error Probable) being about 150 m (492 feet) at maximum range. A few rounds have a smaller 15 KT warhead, and plans are in hand to fit terminally guided submunitions, though the French (undoubtedly wisely) believe in the deterrent power of nuclear weapons. One of the useful adjuncts to Pluton is a real-time reconnaissance update capability using R.20 remotely piloted vehicles carrying IR sensors and able to send back targeting information direct to the computer controlling one or more Plutons. At least ten regiments are in use, with production continuing, and Aérospatiale is now six years into development of a Super Pluton for use late in the decade.

Soviet Union
Since the mid-1950s the Soviets have deployed colossal firepower in the form of free-flight tactical rockets and guided battlefield missiles, using every kind of nuclear, conventional and chemical warhead. All are highly mobile, though the earliest Frog series rockets were clumsily mounted on heavy IS-3 battle tank chassis. Today at least 12,000 amphibious PT-76 chassis are used for various battlefield duties including carrying

several species of Frog. The most important at present is Frog-7, the first to be carried on a high-speed multi-wheel chassis, the excellent ZIL-135, which also carries a crane for reloading. Depending on warhead, Frog-7 has a range of 35-40 miles, and some were fired by Syria against Israel in 1973.

After six years of rumor there is still remarkably little hard fact about the new-generation tactical rocket system, which some Western sources claim is designated SS-21 (by Washington, if not by all NATO) while others continue to describe it as Frog-9. Published descriptions are what one might expect – increased range, inertial guidance by a simple strapdown system and a wide range of payloads – but evidence is lacking.

The large family of SS-1 Scud tactical missiles continues to be deployed, even by WarPac first-line troops, though the individual missiles must all be at least 20 years old. These weapons are carried on tracked or wheeled chassis and have strapdown inertial guidance over ranges up to 175 miles. The larger SS-12 Scaleboard has already been discussed in the first section on strategic weapons. By far the most formidable of all is the relatively new SS-20, also covered in the first section. By March 1983 the number of SS-20 systems deployed was estimated at 380-420 by NATO sources.

Above: Aérospatiale and its predecessors delivered 169,000 SS.11 anti-armor missiles.
Left; Firing an AS.11 (air-launched SS.11) from an Alouette III helicopter in 1974.
Below: While not strictly a missile the soviet RPG-7 anti-tank grenade is a highly effective anti-armor weapon. In the background is a T55 Main Battle Tank.

Top far left: A Soviet AT-3 Sagger anti-tank missile.
Left above: The AT-4 Spigot is one of the new-generation Soviet anti-armor systems.
Left: Frog 7 is one of the later versions of the prolific Soviet Frog rocket family.
Top: A Tupolev Backfire swing-wing long-range strike aircraft with an AS-4 Kitchen supersonic cruise missile.
Above: AGM-45 Shrike was the first dedicated anti-radar missile.

Anti-radar missiles

ARMs grew up as a class in the mid-1960s when it was recognized, first, that ground radars play a key role in air-defense systems, and second, that it is relatively simple to make a missile home automatically onto an operative radar. The first of which details are known was the US Air Force AGM-45 Shrike, based largely on the airframe of the Sparrow AAM but with cropped tailfins and totally different internals. Developed by the Navy and produced by a group headed by TI and Sperry Univac, Shrike is fitted with various passive radio-receiver heads, each tuned to particular hostile radar or communications threats. Early missiles fired in Vietnam turned in one of the worst guided-missile performances on record, but Shrike was progressively

Left: Two Martels carried by a Buccaneer attack aircraft: AS 37 anti-radar (left) and AJ.168 TV-guided (right).
Below: Three TV-guided Martels and the guidance pod carried by a Buccaneer of the RAF.

Above: A Harm together with two AIM-9L Sidewinders on an F-15.
Right: Firing an AGM-88 Harm from a Vought A-7 airplane.

updated and refined until later production – and many refurbished early rounds – offered good reliability and consistent lock-on to the correct target. Shrike is carried by large numbers of USAF, USN, USMC and foreign tactical aircraft, including Wild Weasel platforms assigned specifically to the defense-suppression role.

Standard ARM is a much larger missile based on the Navy's Standard ship-launched SAM and with a firing weight of over 1400 lb. It also involved several industrial teams with different seeker heads, and early production missiles were used in Vietnam from F-105G Wild Weasel aircraft. Production ceased in the late 1970s, but a few of these massive weapons are still used by F-4G and other aircraft.

The Franco-British Martel missile exists in two forms, the French model being the AS.37 anti-radar weapon. Weighing 1168 lb, somewhat lighter than the British TV-guided model, anti-radar Martel is fired over ranges which vary from 18 miles at zero feet to over 37 miles when launched from high altitude, and is carried by many Mirage, Jaguar, Buccaneer and Atlantic aircraft.

As a new-generation missile much better than Shrike or Standard ARM, the AGM-88A Harm (High-speed ARM) was funded chiefly by the Navy from 1972. Weighing about 820 lb, it has a range of over 11 miles and is interfaced with the advanced receiver and analyser systems in the launch platform, which is to be the A-6E, A-7E, F/A-18A, F-4G and EF-111A. Harm is said to have sensitive seekers which can lock-on to emissions too weak or of the wrong character for earlier ARMs.

At this writing there is a sales battle in Britain between Harm and an even newer missile, Alarm, by British Aerospace Dynamics. Alarm (Advanced Lightweight ARM) is an extremely clever weapon which pitches over after release and surveys a wide area of terrain, noting the locations and waveforms of all interesting hostile emissions. It then dives on a selected target; special software (tested in flight trials) enables Alarm to change from one target to another if the first should cease to emit, or, in certain circumstances, to continue to dive on the original emitter location. Alarm is half the weight of Harm, so it can be carried in pairs or triplets on a single pylon.

7. ANTI-TANK MISSILES

Launching a Swingfire heavy anti-tank missile from a British FV.438 APC in Germany in 1974.

Vast numbers of modern missiles are designed for use in land warfare between armies. Some are specifically intended for use against armor, and these form a special class characterized by short range, precision guidance and ability to penetrate the target. On the one hand most armor is slow-moving, unable to dodge and devoid of any kind of countermeasures, either electronic or offensive, by which to destroy or deflect oncoming missiles.

Like all warfare, the conflict between armor and missiles is of a see-saw nature. First armor has the advantage, then more powerful missiles are developed, and these in turn are countered by further improvements in armor. The limits on future development of armor, within reasonable dimensions and weights, are probably more severe than those on future development of missiles, but at present the entire spectrum of armored vehicles, from simple anti-terrorist scout cars to the most costly battle tank, appears to be not only viable but likely to remain so until the end of the 1980s.

At the start of World War II armored vehicles could be engaged by either of two methods (discounting the use of fire, by Molotov cocktails or flamethrowers, both of which are usable by infantry only at very close range). One was the gun, which imparted sufficient velocity for an extremely hard tungsten-carbide bullet or shell core to penetrate the armor solely because of its kinetic energy. The alternative was to use an explosive charge. This can either be powerful enough to disable the tank by brute force, for example, by blowing off the track or turret, or it can be a smaller but more sophisticated charge.

There are two kinds of anti-armor warhead, and both have been much used in modern anti-tank missiles. The HEP (high-explosive, plastic) or in British parlance HESH (high-explosive squash head) contains a substantial charge of violent high explosive, which is detonated by a contact fuze which allows the charge to flatten itself against the armor before detonating. The explosion sends small-amplitude shockwaves through the depth of the armor, and the acceleration at the inner surface is high enough to spall off fragments which wreck the interior and incapacitate the crew. Even more important is the hollow-charge or shaped-charge head, which comprises an HE charge whose forward enclosure has the form of an inverse cone pointing back away from the armor. On being detonated, such a warhead projects a jet of hot gas and liquid metal forward with such velocity that it passes straight through the armor. Thicknesses as great as 30 inches have been pierced by quite modest hollow-charge heads, so most tanks can be defeated even at a very oblique impact angle.

By the end of World War II the hollow-charge warhead had become as important as the previously dominant gun, because it had made possible the German Panzerfaust, the US-developed Bazooka rocket, which was quickly copied and mass-produced by the Germans as the Panzerschreck and the British PIAT (Projector, Infantry, Anti-Tank). All could disable a heavy tank from 100 yards, but even at this distance the chance of a hit was remote. Obviously what was wanted was a small guided missile, and no new invention, as such, was

required. Rocket propulsion, aerodynamic controls actuated under command from signals sent along fine wires and a hollow-charge warhead were all well known before World War II, yet the first anti-tank guided missile did not appear until 1944 and played an insignificant part in the war. Designated X-7 and called Rotkäppchen (Red Riding Hood), it weighed 19.84 lb and carried a shaped-charge head containing 5.5 lb of explosive. It rode on a wing of 23.6-inch span, but rolled continuously as it flew so that the operator could steer it with its single curved arm and tail spoiler. Effective range was three-quarters of a mile, and the front-line troops that received it soon equated one missile with one heavy tank knocked out.

Amazingly, there is no evidence of any effort on the part of the Allies to develop such a weapon after 1945 except for France. This country tested several X-7 missiles in 1945-6, and by 1948 the Arsenal de l'Aéronautique at Châtillon had roughed out the design of the first modern anti-tank missile. After various changes of organization the design team became part of Nord-

Test firing in the USA in 1954 of the pioneer French SS 10 missile.

with modern weapons, a handful of SS.10s absolutely defeated the heavy Egyptian tanks, and this spurred export sales to the point that Nord had sold 29,849 of these effective missiles by 1961. Their main weakness was that they were difficult to steer. For reasons of economy, no autopilot was incorporated, so moving the joystick merely steered the missile in the desired direction. There was nothing precise or positive about it; it took a lot of practice to know just how far to work the control to get a desired result, so the first firing by a new trainee was always a wild series of overcorrections. More often than not the missile hit the ground or a tree, or just missed the target. The importance of SS.10 is that it got everyone started in the anti-tank missile business and also proved that such weapons were useful and could be carried by infantry or grouped in clusters on a Jeep or similar light vehicle.

As SS.10 was originally a product of private industry, the French launched a rival program in the government's DTAT (Directorate of Land Weapons). Called Entac (from Engin Téléguidé Anti-Char, guided weapon, anti-tank) it was significantly smaller and lighter than SS.10, yet had a warhead able to penetrate 25.6 inches of armor after flying over 6500 feet. It was soon obvious that whereas the blunt-tipped wings of SS.10, with a span of almost 30 inches tended to snag on obstructions near the flight path, the swept-back wings of Entac, with a span of only 14¾ inches, flew easily past the same objects and could even zip through light vegetation or outer tree foliage. Even more important, the control system was not of the plain acceleration type but of the velocity type. Instead of manipulating a stick the operator put his thumb in a small pivoted control that was easy to get accustomed to. Moving the control left/right caused an exactly proportionate left or right displacement of the line of flight, and wild overcorrection became a thing of the past. Another advantage was that 10 missiles could be spread out and hidden with loose foliage while connected up to a single operator up to 360 feet away. He could then fire and guide each round in succession. Entac was in production by 1958 and remained so until 1974, by which time 139,417 had been delivered to 14 countries, including the United States where the missile was used in fair numbers as MGM-32.

US work in the field dates from 1951 when the Army Rocket and Guided-Missile Agency, working under the Los Angeles Army Ordnance District, initiated a project designated SSM-A-23 (surface/surface missile, Army, No 23). Industrial effort was centred at Aerophysics Development at Santa Barbara, a subsidiary of Curtiss-Wright, another of whose companies, Utica-Bend, received the August 1956 production contract for $16,565,000, a large order in those days. Named Dart, this missile had wings and spoiler controls similar to those of SS.10 but larger, and at the rear was a set of four tail surfaces. The trouble with Dart was that it was far too big and powerful. It weighed 98.9 lb, and so could not be handled by infantry, and the 40-inch-span wings were too large to pass obstructions easily. Dart was a costly failure, and production was terminated in 1958.

Aviation, which in turn was merged into today's giant nationalized group, Aérospatiale. The missile went into production as the Nord 5203 in October 1955, and was subsequently restyled SS.10 (SS meaning sol/sol, surface-to-surface mission). Weighing 33 lb, SS.10 set the pattern for almost all its successors in having four identical wings spaced in a cruciform arrangement at 90 degrees. Each wing was fitted with a vibrating spoiler control which could make the missile change course. It rolled as it flew, and the operator had to work a small joystick on a control box. Effective range was a mile, and the 11-lb warhead could pierce 16 inches of armor.

SS.10 was made as simple as possible, as befits an issue for a front-line soldier. Each missile cost $950, and the control box (which could be used as often as necessary) only $4500. In the first Israel/Egypt full-scale war

Left above: The US Army called Entac MGM-32; these are on a Hotchkiss field car in 1968.
Left below: Eight Bantam missiles on a Swiss 4×4 Püch-Häflinger truck.
Above: Firing a Bantam from a Swedish MFI.17 lightplane.
Left: A Swedish soldier with an RB 53 Bantam anti-armor missile.

In the British Commonwealth the only anti-tank missile of this early era was even bigger. The monster Malkara, developed in Australia, was as big as a man and weighed 206 lb. It had a range of 7000 feet, but there was no effective way to engage targets at such a distance except under clear desert conditions, and the 57.5 lb warhead was unnecessarily large (more like an anti-ship weapon). Despite this, Malkara completed development in 1956 and went into service with the Hornet scout cars of the British Army. By way of extreme contrast, the Swedes mass-produced a small missile called Bantam, which was made mainly of glass-fibre and weighed only 16¾ lb. Extremely neat, it had folding wings which flicked open as the missile was fired from its carrying box. The 4.2-lb warhead had a shaped charge fired by electrical double-skin fuzing which triggered even at grazing angles greater than 80 degrees. Bantam was a great success, and was not only fired from infantry kits, small 4×4 cross-country cars, and airplanes and helicopters, but was even adopted by

Left: AS.11 being fired from an Alouette II helicopter.
Right: Vigilant anti-tank missile in use by British paras.
Below: An Entac being fired from a quad launcher on an AMX 13 light tank.

Switzerland, which had its own missile called Mosquito (developed by a Swiss/Italian company, Contraves-Oerlikon). Very similar missiles were the West German Cobra and Mamba (which could jump upward from the ground and required no launcher), the Japanese KAM-3D and the British Vigilant, produced by Vickers-Armstrongs (Aircraft) Ltd, which was extremely effective and successful and matched a devastating warhead with a very small missile of only 11-inch span.

In fact, the most successful of all first-generation anti-tank missiles, and still in use in large numbers throughout the world, was a missile deliberately made too heavy for convenient use by infantry. SS.11, the missile developed by Nord after SS.10, weighs 66 lb and flies just twice as fast, at 360 mph instead of 180. Its four swept wings are of only 19.7-inch span, and a completely new feature was that in place of spoilers or pivoted control surfaces, it is steered by four vibrating spoilers which can be controlled to cut across the jet from the sustainer rocket motor which burns for 20 seconds. SS.11 matured in 1956 with various warheads for use against tanks or ships, or for training, and though most were mounted in groups on tanks and other vehicles, some were fired from airplanes and helicopters, and a few even by infantry teams of four, one man carrying three warheads and the others a missile each. The US financed the missile factory at Bourges, where production reached 1500 a month and the total nudged 170,000 by 1979. SS.11 is used by 35 countries; the US Army calls it M22.

By the 1960s this style of wire-guided missile was almost universal for use against armor, but Nord-Aviation led the way again by making the operator's task easier. If a soldier finds it difficult to steer a command-guidance missile in peacetime training, he will find it impossible in war, when he may be in a state of near panic under an armored attack and probably under air attack as well. Judgment of the kind demanded of an airplane pilot had to be eliminated, and the French

Top: A Japanese KAM-3D missile of the JMSDF.
Second above: A Cobra 2000 anti-tank missile of the Federal German army.
Above: MBB of Federal Germany developed Mamba as a successor to Cobra.
Right top: A Japanese KAM-9 anti-armor missile ready to fire.
Right: Vigilant anti-armor missile being fired from a British Ferret scout car.

answer was TCA (Télé-Command Automatique). All today's soldier has to do is look through a bright magnifying optical sight and keep it exactly aligned with the target. The optics of the sight are boresighted exactly parallel to a second set of optics, which receive IR (heat radiation) from a flare on the tail of the missile and focus it on a sensitive localizer receiver. The latter continuously measures the position of the flare in relation to the sightline on the target. Any deviation automatically generates the desired correction signal, which is transmitted along the wires without any action by the operator. Thus the operator merely keeps the sight on target and sees the missile flare held steady dead-center on the sightline. This is the most common method of guidance for anti-armor missiles today, including those fired from helicopters.

The Soviet Union has clearly investigated every kind of anti-tank missile, and the six types of weapon in this class which have so far been identified as being in regular service show many sharply contrasting features. The only unexceptional Soviet anti-tank missile is the oldest, called AT-1 Snapper by NATO. This is very similar to SS.10 in dimensions and concept, but it is considerably heavier at 49 lb and flies further and faster, and its warhead weighs over 11.5 lb. It is instinctive for NATO to underestimate the capability of Soviet weapons, and for many years AT-1 was said to penetrate 13.8 inches of armor – until it was measured by the Israelis at 20.9 inches.

The next system, AT-2 Swatter, represented a quantum jump in technology and even today has no direct counterpart in the West. There are at least three versions, but all have automatic radio-command guidance along the sightline, and doing away with wires enables the missile to fly at 335-365 mph and to be fired over long ranges (limit, 7218 feet) from vehicles and helicopters such as the Ka-25 Hormone and Mi-24 Hind. At least one version uses three different radio frequencies and waveforms to defeat ECM interference, while another version has automatic IR homing on to the warmth of the target. This formidable 55 lb missile is

Left: AT-2 Swatter missile being fired from an Mi-24 Hind D.
Left Below: A Soviet BRDM scout car with five launch tubes for AT-5 Spandrel missiles.
Below: A cutaway AT-3 Sagger Soviet missile and its sight.

carried in groups of four on BRDM scout cars, fired if necessary in rapid succession. Warhead penetration is at least 20 inches.

AT-3 Sagger devastated Israeli armor in the Yom Kippur (October 1973) war, despite the fact that the missile weighs only 24.9 lb. It is a neat weapon, with a conical nose, tubular body, four small wings at the tail and steering by rocket jetevator controls. Many tens of thousands are used by Warsaw Pact armies in at least 18 types of deployment, including two-man infantry teams carrying small cases, a single (reloadable) launcher above the 73-mm gun of the BMP and other APCs, up to eight on the Mi-24 Hind helicopter and 14 (fired six at a time, plus two spare) on the BRDM.

The newest missiles in service are not well known in the West. AT-4 Spigot is a SACLOS (semi-automatic command to line-of-sight) system, with a tube launcher for a missile of 4¾-inch diameter with a range up to 6600 feet at a speed over 400 mph. The SACLOS system is identical in principle with the French TCA, and the same method is used in AT-5 Spandrel, which has a diameter of 6.1 inches and range of 2½ miles. Warhead penetration of AT-5 is known to exceed 27 inches. There are, again, many installations. The BRDM scout cars often have a battery of five AT-5 tubes, plus another five reloads. Alternatively, the AT-4 is often loaded into the two outer tubes, giving a total load of six AT-5s plus eight AT-4s. Both missiles are often seen on small 4×4 Jeep-type cars of airborne units, but neither has been positively identified on the Mi-24 gunship helicopter. On the other hand, the large and very advanced AT-6 Spiral is a basic weapon of the Mi-24 and possibly of the newest battle tanks (M-72 and M-80); it has laser guidance over ranges said to reach 6.2 miles.

The Army's experience with Dart so discouraged the US Defense Department that French weapons were adopted for the 1960s while good US missiles were planned. The first was a very challenging weapon which may prove to have been ahead of its time – because it has suffered prolonged problems – and an indication of the way things must go in the remainder of the 1980s. One basic fault of early anti-tank missiles was that the operator had to have plenty of time and space in which to fire the missile and gather it into his guidance system. If a hostile tank should manage to approach within about 150 yards, it could not be engaged at all, so USAMICOM (USA Missile Command) decided the answer was a gun projectile which could be fired at point-blank range like a low-velocity shell, while at longer ranges it could be guided downstream of the gun barrel to the distant target. The result was MGM-51A Shillelagh, matched to the 152-mm gun of the M551 Sheridan airborne armored vehicle. Weighing 59 lb, Shillelagh is certainly the most complex projectile to be fired from today's guns, yet reliability of the electronic guidance has been satisfactory despite the shock of firing and subsequent rocket flight at very high supersonic speed. IR SACLOS guidance requires only that the operator keep the cross-hairs of his sight lined up on the target, and maximum range is claimed to be no less than 17,060 feet. Ford Aeronutronic (today Ford Aerospace) deli-

Firing a Shillelagh missile from the 152-mm gun of a US Army General Sheridan armored vehicle.

vered 36,000 rounds in 1964-67 at a reported average price of $13,890. Some of these are deployed, with superior results, in M60A2 main battle tanks, which have a slightly different gun.

For use by infantry and light vehicles, the standard USA and USMC anti-tank missile is Dragon M47, of which a reported 250,000 rounds have been delivered (probably finishing in late 1983) by McDonnell Douglas Astronautics and Raytheon. This unusual 13.5-lb missile is popped from its man-portable tube by a solid charge which automatically nullifies the recoil, and is then propelled all the way to the target by the sequential firing of 30 pairs of miniature rocket thrusters arranged down the sides of the 29.3-inch body. Three curved flick-open fins keep the missile rolling rapidly as it flies at up to 560 mph, and the SACLOS guidance steers the missile by selecting particular pairs of thrusters to keep it on course. Kollsman supply the optical/IR tracker and Texas Instruments the night sight, the complete group being attached to successive loaded missile tubes.

By far the most important Western anti-tank missile at present is Tow (acronym, Tube-launched, Optically tracked, Wire-guided), produced entirely by Hughes Aircraft. This was very much the right missile at the right time, and it has been adopted by the USA, USMC, West Germany, Britain, Canada and 23 other countries.

Above: Readying an M47 Dragon missile for an exercise.
Left: M47 Dragon of a US Army airborne division.
Below: Westland Lynx AH.1 helicopter armed with eight TOW missiles.

Firing a TOW missile of the US Marines 2nd Division at Camp
Lejeune, NC, in January 1977.

Top: The giant Malkara anti-tank missile was carried by the British Hornet scout car.
Above: Vigilants being prepared by the UK Duke of Wellington's Regt.
Left: German Cobra 2000 missiles on an exercise.

Tow is a superb exercise in packaging, because few other missiles get so much into a tube of 6-inch diameter, well under 4 feet long (launch weight 46.1 lb in the original model). The missile is popped from its tube by a boost motor and then coasts until well clear of the operator. Then the sustainer ignites and accelerates the missile to 623 mph, almost the speed of sound, so that a range of 12,300 feet can be covered on the missile's own kinetic energy. The usual IR-type SACLOS guidance is used, the missile being steered by four flick-out tail controls driven by pressurized helium gas. The original missile has an 8.6-lb shaped-charge head judged able to penetrate all known armor up to the early 1980s. Tow is not only used on many ground vehicles, but it is also by far the most important Western helicopter missile, arming AH-1 Cobras of various kinds, the British Lynx, Hughes' own 500MD Defender and many other helicopters. Total production in 1983 topped 300,000 rounds, the latest of which is I-Tow (Improved Tow), with various refinements including a long stand-off contact probe. Not yet in production, Tow 2 has the full 6-in diameter maintained right to the nose in a longer body, giving far greater warhead power calculated to defeat all known tanks of the 1980s.

Hughes has also developed a slim, lightweight missile called Tank Breaker to compete for future US Army orders. There is a host of other modern systems, not all of them guided, intended to defeat the awesome and

155

Hellfire laser missiles are the primary weapons of the US Army
Hughes AH-64A Apache helicopter.

Above: Swingfire can be fired from a palletized infantry system, carried by a Land-Rover vehicle.
Left top: Artist's impression of Wasp missiles in action.
Left below: HOT missiles can be carried by the Dauphin helicopter with the Venus system.

growing threat of Warsaw Pact armor. Copperhead could be of very first importance, because this is the first heavy-artillery shell with precision guidance. Developed by Martin Marietta, it is fired from standard 155-mm (6.1-inch) guns and has flick-out wings and tail controls driven according to homing signals from the seeker in the nose, which homes on to any target illuminated by laser light of the correct type. The 140-lb projectile is fired in the general direction of the target and then steered into it by a friendly laser (aimed by an aircraft or front-line soldier) which illuminates the target vehicle. TGSM, Terminally Guided Sub-Munition, is one of several advanced small payloads intended to be dispensed in a cloud from a 'bus' missile such as Lance or a version of the Patriot SAM. Various forms of guidance would steer the munitions into the nearest tanks. Those failing to lock-on to a target would convert themselves into mines lying on the ground.

In Europe Britain's Swingfire, a heavy (59.5-lb) and long-range (2.5-mile) missile with jetevator control is used in large numbers by the British Army and various export customers, and there are new systems under development in France, Italy and Sweden, the latter building an unusual missile (RBS56 Bill) which flies just over the top of the tank and punches it through the relatively thin roof armor with a diagonally directed shaped charge. But the most important missiles are Hot and Milan, both produced by the Franco/German group, Euromissile. Hot is a powerful 55.1-lb weapon for vehicles and aircraft, with the usual IR SACLOS guidance and optional bad-weather and night-sighting systems. Milan is smaller, weighing 14.8 lb and easily handled by infantry. This too has night and bad-weather sights, and production exceeded 150,000 units for many countries by 1983.

Most important of the next generation of weapons from the United States are Hellfire and Wasp; predictably, both break new technological ground and promise remarkable performance. Hellfire is a product of Rockwell International, stemming from the earlier Hornet and with a name derived from HELicopter-Launched FIRE and Forget. The missile is compact, with a caliber of 7 inches and a length of 64 inches; it weighs 95 lb at launch. All initial missiles of this family home automatically onto a target illuminated by laser, but future seeker heads are being studied with IIR (imaging IR), RF (radio frequency, ie, radar), TV, or various dual-mode combinations. Development has gone well so far, demonstrating bullseye accuracy with almost every shot, and Hellfire is scheduled to arm the Hughes AH-64A Apache, UH-60 Black Hawk and probably AH-1T SeaCobra of the Marines. Wasp is the newest Hughes contender and one of the first missiles in the world to have a miniature active radar in the nose working on a millimetric wavelength. Only 60 inches long, and with wings and tail controls spanning 20 inches when unfolded, Wasp is carried in multiple-tube launchers by all kinds of aircraft from the F-15 and F-16 down. One of the obvious problems has been to devise a method for assigning a cloud of 16 Wasps to 16 different tanks. This problem is solved as long as there are more targets than missiles.

159

8. ANTI-SHIP SYSTEMS

An Israeli-developed Gabriel missile being fired from a fast patrol boat.

The very first guided missiles ever tested were designed for use against ships. As early as October 1914 Dr Wilhelm von Siemens suggested that his company should build radio-controlled glider bombs, and from January 1915 the vast Siemens-Schuckert Werke tested large numbers of glide-missiles, some biplanes and some monoplanes, from airships and heavy bombers of the Imperial German Navy. Some had radio guidance but most pioneered the technique of guidance via electrical signals passed along wires trailed behind the missile. Some SSW missiles released torpedoes at the target while others had integral warheads.

In World War II German designers led the way again with six major families of anti-ship missile, the largest of which were the Mistel (Mistletoe) composite aircraft consisting of Ju 88s rebuilt with a gigantic warhead in place of the crew compartment, guided by radio from a fighter which flew to the target area actually mounted above the pilotless bomber. A few of these were tested against old warships, and one with a 7716-lb warhead almost blew the battleship *Océan* in two. Operational Mistels were used against Allied ships in the English Channel in the summer of 1944.

More effective were the purpose-designed missiles, of which two types were delivered in large numbers for air launch by the Luftwaffe. Largest was the FX, or Fritz X, which comprised a giant bomb with four cruciform wings and a complicated tail incorporating radio guidance and control vanes. Weighing 3461 lb, it was a

**Above: An early McDonnell F-4B Phantom fires a Bullpup ASM.
Left: The Swedish AJ37 Viggen supersonic airplane carries the
big RB 04E anti-ship missile.**

powerful weapon, and from 29 August 1943 it destroyed many Allied ships. On 9 September the Italian
fleet sailed from La Spezia to join the Allies and six Do
217K-2 bombers caused havoc, sinking the battleship
Roma and leaving her sister *Italia* with her decks awash.
Many other ships fell victim to FX, but the smaller
2304-lb Hs 293 was used in even larger numbers, some
2300 being launched. An Hs 293 sank the escort HMS
Egret on 27 August 1943, the first ship ever destroyed
by a missile. Hs 293 was propelled by an underslung
rocket over ranges up to 11 miles, and carried bright tail
flares to help the operator steer it into its target by radio.

Precisely the same scheme was used in the American
Bullpup, the first post-war anti-ship missile to be produced in quantity. This was odd, because long before
Bullpup US organizations, led by the National Bureau
of Standards, had created an 1880-lb missile, Bat,
which had the much more advanced guidance of a small
radar in its nose, homing on the reflections from the
target ship. Bats were hung under the wings of US Navy
PB4Y-2 Privateer patrol bombers, and from May 1945
they sank many Japanese ships, including a destroyer
from a range of 20 miles. Active-radar guidance was
tried in several US Navy missiles of the 1950s which
failed to leave the laboratory, and by comparison Bullpup was crude, being described as 'a 250-lb bomb with

a guidance section on the front and a rocket at the back.'
Produced by Martin's Orlando division, Bullpup entered service in 1959 and subsequently 22,100 of the
original model were delivered, plus over 8000 made in
Europe, for use by many aircraft ranging from the A-4
and F-104 to the P-3 Orion. Bullpup B (now designated
AGM-12C) is larger, with a 1000-lb bomb as basis.

Bullpup was better than 'iron bombs' in the days
when aircraft lacked today's precision microelectronic
weapon-aiming systems, but it suffered from many
faults. There are obvious disadvantages in an operator
in the launch aircraft, or another flying nearby, having
to use a miniature joystick to keep bright flares on the
missile lined up exactly with the target until it hits. It
means the aircraft has to stay quite near the target and
avoid violent maneuvers throughout the time the missile is heading toward the target; in fact, to give the
guidance operator a good view, the aircraft is almost
forced to circle round the target, making the task of
anti-aircraft gunners almost simple. Much thought was
devoted to trying to find a better system, some alterna-
tives being tested on Bullpups.

Remarkably, the Swedes had produced a better missile and got it into service earlier. Using the active-radar
homing first used on the Bat, groups of Swedish companies, plus a British rocket-motor supplier, began
work in 1949 on RB 04 (RB for 'robot,' Swedish name
for missiles) and got it into service in 1958. The original
version weighed 1323 lb and had four control fins at the
front (like Bullpup) but a large delta wing at the rear
with twin fins at the tips. All versions have a 661-lb
warhead and can fly up to 20 miles, homing automati-

Below: Launch of AS 30L laser-guided ASM from a French AF Jaguar with Martin Marietta Atlis 2 pod.
Below: AS 30L test launch against a concrete target at the Center d'Essais des Landes (France).

cally on the target. Thus, the RB 04 has from the start been a so-called 'launch and leave' missile, something many users are still crying out for. As soon as the launch aircraft (now the AJ37A Viggen) has sighted the distant target on radar, it can cue the missile's own radar to lock-on and release the missile. It can then immediately turn away, leaving the missile to home on the target by itself.

As described earlier, Nord-Aviation in France developed small wire-guided anti-tank missiles in the 1950s, and from these stemmed larger weapons for use against different targets, including ships. One is AS.12, weighing 170 lb and with a 62.6-lb warhead designed to penetrate 1.57 inches of armor before exploding. This missile has been carried by seven types of airplane and seven types of helicopter, with deliveries amounting to over 8400. AS.12s fired by Royal Navy Wasp helicopters sank the Argentine submarine *Santa Fé* at Grytvyken on 25 April 1982. SS.12M is a version carried by FPBs and other small naval craft. AS.20 and AS.30 are faster air-launched missiles which can be used against any kind of surface target, not just ships. The latest AS.30L version, weighing over 1200 lb, has laser guidance and homes from up to 7 miles on any target illuminated by a laser (carried in the launch aircraft or aimed by any other friendly platform).

For sheer variety and weapon size, no anti-ship missiles can rival those of the Soviet Union. Many types have been developed, some for air launching and others for use by surface ships and submarines. All the early versions had a basic configuration resembling that of a small airplane rather than the more modern shapes of supersonic missiles, though in general they carried devastating warheads. As with other weapons, NATO assigns its own code names to Soviet anti-ship missiles. The first, SS-N-1 Scrubber (also called Strela) remains an enigma, but it was certainly a large airplane-type weapon carried in groups in hangars at each end of *Krupny* Class destroyers and fired from large rail launchers which could be turned to point in the desired direction. Much better known is SS-N-2 Styx, which has a span of 9 feet 2 inches, length of almost 21 feet and weight of 5550 lb complete with underslung booster rocket fired at launch. Because of its thick fuselage, this weapon flies at not more than about 600 mph on the thrust of a cruise rocket in the tail. Most of the mission is flown on autopilot, though some can have radio-command guidance. Nearing the target, which can be up to 26 miles distant, the terminal guidance is switched on, and this can be either active-radar homing or IR (heat) homing. Styx missiles have been exported in large numbers, and many hundreds have been made in China.

With a warhead weighing 882 lb, Styx can destroy all but the largest warships and cause severe damage to the latter. One of the significant things about it was that it was packaged into aluminium box launchers which were then installed on small FPBs like the *Komar* (two missiles) and *Osa* (four). These extremely small, fast and agile boats have been exported in dozens and even today are used by 22 or 23 navies. Yet the threat they posed to the established naval powers was ignored, until

Left: N-2 Styx missile being hoisted aboard.
Above: Launch of an N-2 Styx anti-ship missile from a small Osa class boat.
Below: Kresta 1 class cruiser of the Soviet navy with SSM (N-3 Shaddock) launch tubes.

on 21 October 1967 the Israeli destroyer *Eilat* exploded in a cloud of smoke and the debris sank into the Mediterranean. She had been struck by two Styx missiles fired by a small *Osa* boat that had not even bothered to leave Alexandria Harbor. There was nothing new about the event; such things had happened back in the 1950s when Styx was under development. But it caused ripples of shock in Western naval staffs and triggered off a rash of new missile projects, some to shoot down weapons like Styx and others to copy it.

While Styx itself was improved into SS-N-2 (Mod), a new-technology version also originally called SS-N-11 – a far bigger and more powerful weapon – entered service from 1958. Developed in the same 1951-56 timeframe as Styx, SS-N-3 Shaddock is reckoned to weigh over 9900 lb and to be 42 feet 8 inches long. It is fired from any of several forms of container/launcher by two rocket-boost motors, and then flies at supersonic speed for up to 286 miles. The warhead is universally judged to be nuclear, in the 350- or 800-kiloton range, and guidance has always been thought to need radio assistance from a ship-based helicopter (Ka-25 Hormone) or strategic shore-based airplane (Tu-142 Bear). Scoop Pair radars are used on Shaddock-armed surface ships for initial guidance, and most Western observers believe that active-radar terminal homing is used. The chief surprise about this formidable weapon was the extent and urgency with which it was fitted aboard submarines. Starting with converted *Whiskey* Class vessels, Shaddock became primary armament of the *Juliet*, *Echo I* and *Echo II* boats, the 27 E-IIs having four twin launchers lying flush with the top decking of the hull when not in use. Other installations are on *Kynda* and *Kresta I* cruisers. A shore-defense version mounted on giant transporter/erector/launcher vehicles is called

Below: The Kynda class cruisers also carry the big N-3 Shaddock cruise missiles.
Right: Krivak class gas-turbine warships have SS-N-14 missiles whose chief mission is anti-submarine warfare.
Far right: The extremely successful Nanuchka class vessels have a wealth of weapons and sensors including tubes for N-9 missiles.

Sepal, and is fired straight from its tube.

SS-N-7 is a newer weapon, and the first Soviet missile designed for launch from a submerged submarine. Not much is known about it, except that it arms the extremely fast nuclear *Charlie* Class submarines. Figures accepted by the Pentagon include a missile length of 22 feet and weight of total warhead of 1100 lb (nuclear, 200 kilotons), and a range of 6 to 35 miles at Mach 0.95. First seen in 1968, this missile was developed in the era when it was realized that reaching a heavily-armed warship bristling with radar required that a missile, like an airplane, come in as low above the water as possible. Thus today the 'sea skimmer' kind of anti-ship missile forms a totally distinct class. Previous anti-ship missiles with integral warheads might have been used against land targets, but the sea skimmers are held by radio altimeter and autopilot at a very low height above the sea which would be impractical over any landscape. This in effect means that the modern anti-ship missile needs to 'think' in only a single plane, its seeker in the nose searching left/right but ignoring up/down scans. The Soviet N-7 was not the first such missile, but it was the first to combine sea skimming with submerged launch.

The newest Soviet naval cruise missiles are N-9 Siren, N-12 Sandbox and N-14 Silex. N-9 is 30 feet long and flies up to 68 miles at Mach 0.8 from a submerged launch (*Papa* submarines) or surface launch (*Nanuchka*

Left: Newest Soviet destroyers, the Sovremennyi class have many weapons including N-9 anti-ship missile tubes.
Far left below: Kashin class destroyers have a double-ended (two twin) SAM system and multi-barrel ASW launchers.
Below: Weapons of the Kiev class include: A, MBU-4500 (ASW); B, SUW-N-1 (ASW); C, twin 76-mm guns; D, 12 SS-N-12 (anti-ship missiles); E, SA-N-3 (SAMs); F, 23-mm Gatling (close-in anti-missile or AAA); and G, retractable SA-N-4 (SAMs).

misile boats), carrying 200-kiloton nuclear or 1100 lb conventional warhead. N-12 is an updated Shaddock, with a range up to 350 miles at Mach 2.5 carrying a 350-kiloton or large HE warhead and sea-skimming the final 10 miles. N-14 is both an anti-ship and ASW (anti-submarine-warfare) missile, with a nuclear (five-kiloton) acoustic homing torpedo as payload in the ASW role. Maximum range is estimated at 34 miles.

In addition to these surface-launched weapons, the Soviets have produced large numbers of giant cruise missiles launched by bombers. Some can also be used against land targets, although most have been designed primarily to knock out the largest units of Western navies. AS-1 Kennel resembled a scaled-down MiG-17 fighter, which is not surprising because, like several of its successors, it was designed by the MiG aircraft bureau. Tu-16 Badger bombers appeared with one of these 6600-lb missiles under each wing in 1957, and many were supplied to Egypt, China and Indonesia with the same aircraft. This turbojet missile was replaced in 1965 by AS-5 Kelt with rocket propulsion in a different airframe resembling a baby MiG-19. Most AS-5s have a 1-ton warhead and an active radar in the nose for homing on ships. One version has passive homing for use against operating radars, and at least two struck Israeli radar stations on land in October 1973 when about 15 were successfully intercepted by Israeli fighters, despite their speed of Mach 1.2. Maximum range of AS-5, which is still in use, is 143 miles, or less in an all-low-level mission.

AS-2 Kipper was larger, so that only a single weapon could be carried under the centerline of a Badger-C bomber. Powered by a turbojet underslung at the rear, AS-2 resembles the defunct USAF Hound Dog but has active-radar terminal homing after flying up to 115 miles at Mach 1.4; the warhead is reckoned to be 1-ton HE. The biggest air-launched missile of all, and used as much against cities as against ships (such as the largest carriers), AS-3 Kangaroo is carried only by the monster Bear-B and Bear-C airplanes. Estimated to weigh just over 22,000 lb, it is propelled by a turbojet at Mach 1.8 for up to 400 miles. The exact methods of guidance are not known, but the 800-kiloton nuclear warhead could devastate an entire battle fleet.

AS-4 Kitchen is an extremely fast (Mach 3.5, 2450 mph) missile carried under both the Tu-22 Blinder and Tu-26 Backfire-B bombers. Carrying a 1-ton HE or 350-kiloton nuclear warhead, it can fly 190 miles at low level, or as far as 290 miles at high altitude, and different versions have inertial mid-course guidance and either active- or passive-radar terminal homing. The active type uses its own radar to find the biggest ship, while the passive type homes automatically onto the most suitable operating radar toward which it is heading. AS-6 Kingfish is very similar but slightly smaller, and it is used by both the Backfire and a newly modified Badger-H version of the Tu-16.

Discussion thus far shows some of the basic factors affecting the design of anti-ship missiles. Ships are relatively large masses of metal and form almost perfect targets for an active-radar homing missile, but warships are specially equipped with ESM (electronic support

Above: Artist's impression of Tupolev Backfire bomber. Right: An AS-1 Kennel, the first Soviet air-to-ship missile.

measures) which give instant warning of illumination by a hostile radar. Thus the missile must either avoid using radar or be so small and/or agile that it is very difficult to shoot down. Sea skimming makes it very difficult to detect on the ship's own radars, but at the same time it prevents the missile from dodging defensive fire except in a limited side-to-side movement. An alternative method is to fly as far as possible toward the ship just above the waves and then suddenly climb and dive on the ship from above. This gives a better view of the ship to the missile radar (or other homing system), is likely to confuse the defenses in the last few vital seconds before impact and also makes it easier for the warhead to penetrate inside the ship before exploding.

Below: Launch of a Gabriel, the Israeli anti-ship missile.

Probably the first anti-ship missile to pop-up and dive on its target was the Swedish RB 315, a 3090 lb cruciform-winged weapon propelled by an internal resonant-duct ramjet at 590 mph for ranges up to 9½ miles. This hefty weapon armed two destroyers in 1955-65 and had radio-command guidance which restricted it to line-of-sight ranges. The next Swedish missile in this class, RB 08A, was a sea-skimming airplane-type weapon weighing 2679 lb and flying on autopilot and height lock until its own active radar was switched on near the target. This enabled it to operate far over the visual horizon, the range being up to 155 miles. Norway developed Penguin, a 750-lb missile fired from small boats like FPBs over ranges up to 18 miles. This weapon emits no tell-tale signals, but instead has inertial mid-course guidance and then homes either on the target's heat (IR) or on its radio or radar signals (passive homing). In complete contrast, the Italian Sea Killer missiles are dependent on giveaway radar and radio signals from the launching vessel, which are intended to steer this 661-lb rocket into the target at distances up to 15½ miles. Another weapon system of the early 1960s is the Israeli Gabriel, the first version of which weighs only 882 lb yet carries a 331-lb warhead and can fly over 13 miles. Most versions are well equipped to deal heavy blows to sophisticated warships well endowed with defenses, including advanced ECM (electronic countermeasures), and in fact Gabriel can home automatically on hostile jamming. In its terminal phase Gabriel sinks until its wings almost brush the waves and uses semi-active homing onto a target illuminated by radar in the parent ship.

Left top: Launch of a Harpoon cruise missile from PCH-1 hydrofoil *High Point* in 1974.
Left: Launch of a BGM-84 Harpoon from a US Navy surface ship in December 1972.
Top: Launch from a US Navy P-3 Orion of AGM-84A Harpoon.
Above: Cutaway drawing of the McDonnell Douglas Harpoon missile.

All these weapons were in use before the *Eilat* was sunk in 1967. Then came a sudden spurt of anti-ship missile work and defense countermeasures. By far the most important project in the United States is Harpoon, which began in 1968 as an air-launched anti-ship missile and later grew into an extremely versatile family of basically similar missiles launched from aircraft, surface ships and submarines (designations AGM-84A for air and RGM-84A for ship/sub). After RFPs (requests for proposals) had been answered, McDonnell Douglas Astronautics was picked in June 1971 as prime contractor. Since then, after the usual early technical snags, Harpoon has been a most successful program.

Harpoon is not like a small airplane but has cruciform wings amidships and tail controls, though to get the desired very long range it was decided to power it by a small turbojet (Teledyne CAE J402 of 660-lb thrust), which draws kerosene from a tank amidships and gives a range of over 57 miles, enough to engage targets far over the horizon from any low-level launch platform. AGM-84A weighs 1160 lb at launch and carries an NWC (Naval Weapons Center) 500-lb penetration/ blast type warhead which can make a mess of any ship and disable any smaller than a large cruiser or carrier. Mid-course guidance is by a strapdown inertial platform, eliminating any need for a link with the parent aircraft. Upon nearing the target, the Texas Instruments frequency-agile radar is switched on and this locks-on to the target. In the final few seconds the pop-up maneuver is usually commanded (it is optional), the radar maintaining its lock on the target throughout. The missile remains under jet power and has high maneuverability, such that no ship could evade it.

For ship launch RGM-84A has to have a tandem-booster rocket, which increases launch weight to 1470 lb. A very important feature is that the Harpoon was made compatible with existing standard ship launchers and handling systems, already in use for Tartar, Standard and Asroc. For ships not equipped with such launchers, Harpoon can be packaged into a tubular canister with its wings and tail surfaces folded; four canisters are then mounted on a fixed ramp in the PHM (Patrol Hydrofoil Missile) craft and other small surface vessels. For use from submarines the missile is enclosed in a watertight canister and fired like a torpedo from regular 21-inch tubes. Sub-launched Harpoon is being installed in all SSN attack submarines, and in a non-standard British launch capsule from similar submarines of the Royal Navy. On breaking the surface at a near-vertical angle, the capsule is blown open and the missile boost motor fires, the rest of the mission being as for other versions.

After a delayed start McDonnell Douglas managed to deliver 315 of all versions in 1976, and by 1983 the total exceeded 2100. By 1988 the total should be well beyond 5000. So far, 12 foreign navies are known to have placed orders.

The other leading anti-ship missile in the West, the French Exocet, gained worldwide publicity in 1982 as the result of its supposed destruction of the British destroyer HMS *Sheffield*, an advanced modern warship, and subsequently the large container ship *Atlantic Conveyor*. Without wishing to detract from the Exocet's no doubt deserved success, the actual performance of the missile was not impressive; the two known to have struck British warships in the Falklands war might as well have had concrete warheads, because neither exploded!

Below: Firing an Asroc ASW missile from USS Brooke (FFG-1) in 1969.

Loading a Subroc ASW missile down the forward hatch of a US Navy *Permit* class submarine.

182

Loading a Subroc ASW missile down the forward hatch of a US Navy *Permit* class submarine.

Top left: Cat launch of a French Super Etendard with AM 39 Exocet.
Left below: Four-shot MM 40 Exocet coastal battery.
Above: Damage-control team of HMS *Glamorgan*, Falklands, 11 June 1982.

Starting immediately after the *Eilat* sinking, Nord-Aviation (today Aérospatiale) was able to draw on long experience of previous missiles and finally selected the cruciform-wing layout of AS.30, the guidance of the German Kormoran (described later) and rocket propulsion based on that of air-ground Martel. Though it reduces maximum range, rocket propulsion was adopted because it enables missiles to be fired at only a few seconds' notice (though Aérospatiale are overstating their case when they claim turbojet missiles like Harpoon need about a minute's warning). Another reason is that the French company claims turbojet propulsion necessarily involves large air inlets 'projecting like horns, making large targets for the defending radars.' A walk around Harpoon does not reveal any projecting inlet; the air for the engine is drawn in via a flush inlet on the underside between the two lower wings, and the radar cross-section is not enhanced at all by the inlet's presence. Possibly rocket propulsion reduces cost, and it may be slightly less prone to failure than a complicated turbojet, but there is no clear advantage and certainly McDonnell Douglas and the US Navy are convinced Harpoon has the preferred propulsion.

The first version of Exocet to go into service was MM.38, for use from surface ships including air-cushion vehicles (surface-effect craft) and hydrofoils. This version is launched from a large aluminium box container with a loaded weight of 3858 lb. In most warships four are arranged either in a compact group or

in a fan, each launcher being fixed in azimuth and elevation. The missile is blasted out on its boost motor, weighing 1620 lb at the moment of firing, and flies on inertial guidance at a height of typically 8 feet at a speed of Mach 0.93 (about 710 mph). Maximum range is 26 miles, and in the final 35 seconds (about 7-8 miles) the X-band monopulse homing radar is switched on to search about two axes for the biggest target only. The warhead is designed to penetrate light armor at an angle of 70 degrees from normal (ie, a 20-degree glancing blow) and detonate the 364-lb HE and steel-block charge inside the hull. The Royal Navy fired the 101st production round against HMS *Undaunted*, an old destroyer, and almost blew the ship in half.

Next came AM.39 for air launch. Some of the earliest AM.39s were successfully fired in the hover by Super Frelon helicopters, but the main carrier airplane is the Super Etendard carrier-based attack aircraft, which normally carries one AM.39 balanced by a drop tank. AM.39 has different boost and sustain rocket motors, giving ranges (depending on launch height and speed) up to 44 miles. The missile is shorter than MM.38 (15 feet 4½ inches compared with 17 feet 1 inch) and weighs only 1444 lb. Full information on the Argentine Navy Super Etendard attacks on the British task force may never be forthcoming, but certainly the Exocet that struck *Sheffield* did not explode. The ship was lost to uncontrollable fires, and this was entirely due to shortcomings in the design of the ship which suffocated crew below decks and prevented pumping of water. The two missiles which hit the container ship exploded, but that was a merchant ship not designed to withstand missile attack. In contrast, HMS *Sheffield* was designed to detect both aircraft and missiles on her own radar and also to sound an alarm should the ship be illuminated by hostile radars. It must have been thus illuminated once

Westland Lynx HAS.2 of the Royal Navy with Sea Skua missiles.

Left top: Aérospatiale AS 15TT missiles (mock-ups) on a
Dauphin helicopter.
Left below: Israeli F-4E Phantom with the new Gabriel III/AS.
Above: An RAF Buccaneer bomber with four Sea Eagle cruise
missiles.

(at least) by the Super Etendard and subsequently by
the radar of the Exocet itself, yet no warning appears to
have been given. No explanation of the ship's apparent
failure to detect either the attack or the radar emissions
has yet been given.

In the final Exocet firing, in the closing phase of the
campaign at Port Stanley, the missile was an MM.38
whose launch box was one of a pair welded to a large
land trailer. It was fired at 20 miles' range against HMS
Glamorgan, using land radars, and this time the ship
was alert. She was maneuvered stern-on to the missile,
which was plotted on radar, and though a 6200-ton ship
cannot dodge, it can fight back, and in the closing
seconds it grazed the Exocet with a Seaslug (a primitive
SAM). This may have affected the fuzing, because
though the Exocet ricocheted off the stern, passed
through the helicopter hangar and detonated in the
crowded galley and mess-deck, the main warhead did
not explode. The minor explosion ruptured the heavy
steel-block head, causing 13 casualties, but the ship was
hardly damaged.

Next Aérospatiale produced the MM.40 as a second-
generation ship-launched version, with a new boost/
sustain rocket motor giving a sustainer burn time increased from 93 to 220 seconds. The homing head has a
wider search/acquisition gate angle, and in its final run
to the target MM.40 sinks almost to the wavetops to hit
even the smallest target. Though its launch weight is

increased to 1874 lb, the use of a new glass-fiber tube
launcher reduces the weight of the loaded launcher
dramatically, to only 2535 lb. The MM.40 folds its
wings and control fins to fit the tube, and four launchers
can be accommodated in the space needed by a single
MM.38 box. Newest of the Exocets is SM.39, still
under development, in which a missile similar to
AM.39, but with folding wings and fins, is fired like a
torpedo inside a sealed capsule. The latter has its own
underwater rocket propulsion, with jet-spoiler steering, and on breaking the surface the missile is ejected in
the same way as the sub-launched Harpoon. Range of
SM.39 will be 31 miles.

So far Aérospatiale has sold Exocet in all its versions
to the French Navy and in other forms to 26 foreign
customers (probably more, in view of the current interest). Thus it is certainly the top anti-ship missile in
terms of number of users.

France has a half-share in another missile, Otomat, a
name derived from those of the two partners, Oto
Melara of Italy and Matra of France. Otomat resembles
Harpoon, in that after being fired from a ship or hydrofoil by means of boost rockets burning for four seconds,
the cruise propulsion is by a turbojet. The engine is a
Turboméca Arbizon (French) of 836-lb thrust, and it is
fed by four inlets at the roots of the wings which probably do increase the radar signature. Range in the Mk 1
missile is 37 miles and the Mk 2 extends this to a
claimed figure in excess of 62 miles. Mk 1 cruises on
radio height-lock until at 7½ miles the active radar
locks-on; the final attack is a pop-up and swoop from a
peak of 574 feet, resembling Harpoon. Mk 2 skims the
sea all the way, using single-axis radar. Otomat has been
exported to at least eight countries.

Martel was a joint ASM project between British Aerospace and Aérospatiale, and while one version (AS.37) homes on radars, the British AJ.168 model has TV guidance and can be steered into any target chosen by the crew of the launch aircraft. Another ASM is the German Kormoran, developed by MBB with Aérospatiale aerodynamics and using French rocket motors and radio altimeter. Weighing 1323 lb, it has a range up to 23 miles and skims the sea lower and lower until impact. The 364-lb warhead contains 16 radial charges which project linear fragments with a velocity high enough to penetrate (on average) seven ship bulkheads. The German Marineflieger F-104G Starfighters carried two Kormorans, and the new Tornados carry from four to eight. In a smaller size class come the French Aérospatiale AS.15TT, carried by the Dauphin helicopter and guided by helicopter radar up to 9.3 miles (developed with Saudi money, this missile may be ready in 1984-5), and the British Aerospace Sea Skua, which is matched to the Lynx helicopter and is a more powerful 320-lb weapon. In the Falklands war Sea Skua was hastily cleared for action, and in the worst possible blizzard conditions scored seven direct hits from seven firings.

Newest of the anti-ship missiles in use are the British Aerospace Sea Eagle and Israeli Gabriel IIIA/S. The latter is an air-launched version of Gabriel, upgraded in weight to 1323 lb and with range of 37 miles with a 331-lb warhead. It is fired from A-4 and F-4 aircraft of the Israeli AF. Sea Eagle is in the same size class but has a Microturbo TRI.60 (French) turbojet of 787-lb thrust, to give a range of 'several tens of miles.' It is claimed to have the most advanced on-board systems and active-radar homing head of any anti-ship weapon. Buccaneer and Tornado aircraft normally carry four and Sea Harriers two. Yet another new European missile is the Swedish RBS 15F, again with a Microturbo jet engine and launch weight of 1318 lb; a fire-and-forget weapon, it is launched from small ships or from the supersonic Viggen aircraft.

For the future there is little doubt most design teams are looking at integral ramjets, hybrid (rocket/ramjet) engines or ducted rockets in order to combine highly supersonic speed with increased range. Doubled speed means four times the drag, and thus four times the propulsive thrust. Air-breathing propulsion is the only way to get the range, because it uses oxygen from the atmosphere, which in a rocket has to be carried on board. The typical picture of a 1990s anti-ship missile is a very slim streamlined body (size depending on envisaged size of target, but probably well over 20 feet long) with cruciform wings and flush recessed air inlets to the internal combustion chamber(s) surrounded by solid fuel. The first part of the charge would be packaged to form a fast-burning rocket for high launch acceleration

to supersonic speed. Thereafter the reduced thrust of the sustainer would maintain speed at perhaps Mach 2 (say, 1500 mph) all the way to the target. Higher speed means the defenses have less time to take action to protect the ship.

Defenses may take the form of a rapid-reaction precision missile such as Seawolf (the only one in the world in wide use that was specifically designed to kill anti-ship missiles), or rapid-fire guns, such as the US Navy's Phalanx, or even a system such as the French Javelot which blasts off hundreds of small spin-stabilized high-velocity rockets in tightly spaced groups about 0.2 second apart.

Launch of the new Swedish RBS.15 anti-ship missile.

INDEX

PIAT (Protector, Infantry, Anti-Tank) 138
Panzerfaust, panzerschreck 138
Papa Class submarines, Soviet 171
Peenemunde missile base 9, 10, 12-13, 23, 32
penetration aids 18
Philco Corporation 115

Qian Xvesen 64

RCA 53, 101
Raborn, Admiral 32
radar
    automatic phased array 18
    in defense 54, 95, 97, 104-6, 109, 113, 114-17, 118, 131
    missile homing 163, 166, 167, 172, 175, 177, 188
    missile tracking 104
    missile site radar (MSR) 18, 54, 55
    perimeter acquisition radar (PAR) 18, 54
    semi-active homing 109, 114, 115, 116-17, 118, 120
    SPY-1A phased array radar 101, 104
Raytheon 109, 117, 151
Redoutable, French SSBN 62
Renown, HMS 58, 59
Repulse, HMS 58, 59
Resolution, HMS 58, 58, 60
Revenge, HMS 58
rocket artillery, Soviet 67, 128
    Frog series 80, 128-9
    Frog-7 81, 129, 130
    Frog-9 (SS-21) 79, 80, 129
rockets 125
    A1, A2, A3 (German, later V-2) 12
    A9, A10 13
    V-1 (Fi 103) 9, 10, 10, 12, 22, 125
    V-2 (A-4) 9, 10, 10-11, 12-13, 13, 23, 27, 28, 31, 32, 64, 80, 82, 125
    French Javelot system 188
    Henschel Hs 293 12
    history of 8-9
Rockwell International 159
Royal Air Force 60, 104, 109, 111
    Bomber Command 12, 31, 37, 58, 60
Royal Navy 58, 60, 104-5, 113, 166

SEP company 128
satellite early warning 95
satellite navigation systems 16
satellite reconnaissance 37, 39
Saturn V space vehicle 12
Schneider, General 12
Schriever, General Bernard A 28
Scicon company 128
Sheffield, HMS 178, 183
Siemens, Dr Wilhelm von 162
single integrated options plan (United States) 38, 42
solid propellants 27, 32, 35, 37, 40, 41, 44, 50, 53, 64, 65, 70, 71, 73, 75, 79, 83, 89, 94
Soviet Union
    air defenses 95-7, 105, 116
    anti-ship missiles of 116-7, 171-2
    Chinese danger to 75
    Ground Forces 68-9, 80
    Long Range Aviation Group 82
    military threat of 22, 27, 37, 38, 39, 54
    MIRV first deployed 14
    missile deployment by 90-91, 93-4
    missile development in 12, 13, 29, 39, 68-77, 79-80, 82-4
    National Air Defense Force (PVO Strany) 95
    Navy 74-5, 76-7, 84, 94
    Regional Nuclear Force 75-6, 79
    space program 12, 32, 69, 70, 71, 87, 89
    strategic requirements 68, 70, 71-7,

80, 84, 89, 91
    Strategic Rocket Forces 50, 69-71, 74, 77, 79, 83, 84, 87, 90-91, 97
space exploration 8, 10, 12, 32, 69, 70, 71, 87, 89
Sperry Corporation 128
Sperry Univac 131
Sputnik program 12, 32, 69
Stonewall Jackson, USS, SSBN 33
Strategic Arms Limitation Talks (SALT)
    SALT I 39, 41, 79, 94
    SALT II 79-80, 87, 91
    Anti-Ballistic Missile Treaty 55, 76, 95, 96, 97
strategic bombers 16, 70, 96-7
    British 37, 58, 60
    Chinese 64
    French 61
    Soviet 50, 69, 70, 79
    US 22, 23, 24, 27, 28, 37, 38, 39, 40, 43, 70, 71, 73, 75, 95, 96-7
submarines, nuclear, attack 48, 178
submarines, nuclear missiles carrying (SSBN)
    British 37, 58, 60, 178
    Chinese 65
    French 61, 62, 63, 64
    Soviet 71, 74, 76-7, 80, 82, 84, 92-3, 93-5, 167, 171
    US 14, 16, 23, 27, 32, 32-3, 38, 39, 44, 50, 73, 74, 77
Sweden, missile development in 163, 166, 175, 188

Terminally Guided Sub-Munitions (TGSM) 127, 128, 159
thermonuclear bomb 28
Tsiolkovsky, K E 8
Tunny, USS, submarine 24
Turkish Air Force 29, 62
Typhoon Class SSBN, USN 80, 89, 94

Undersea Long Range Missile System 44
United States
    air defense 95, 100-101, 104, 107, 109, 114-16
    and Cuban missile crisis 72
    and SLBMs 14, 16, 32, 35
    anti-ship missiles of 163, 177-8
    anti-tank missiles 139, 148, 151, 155
    anti-submarine warfare capability 16
    ballistic missile defense system 18
    defense strategy 23, 27, 37-9, 41-2, 71, 95
    Joint Strategic Planning Staff 38
    MIRV deployed 14
    Missile Command (USAMICOM) 148
    missile development in 12, 13, 14, 22-4, 27-9, 31-3, 35, 37-44, 48, 50, 53-5, 70, 91, 114-18, 120-21
    Strategic Air Command established 22
    Strategic missile forces 22-55
    tactical land missiles 125, 127-8
    United States Air Force 23, 24, 27, 28-9, 31, 35, 38, 39, 53, 60, 112, 120, 134
        Strategic Air Command 22, 28, 29, 31, 37, 38, 41, 43, 44
        Tactical Air Command 24, 125
    United States Army 27, 28-9, 31, 32, 35, 37, 50, 53, 54, 104, 109, 111, 125, 127
    United States Army Air Force 22
    United States Navy 23, 24, 27, 28, 31, 32, 35, 38, 39, 100-101, 104, 120, 125, 131, 134

Vickers-Armstrong (Aircraft) Ltd 144
Vostok space launcher 72
Vought Corporation 127
    see also Chance Vought

War of 1812 8
West German Luftwaffe 37

Western Electric and Bell Telephones 53, 54, 104
Whiskey Class Submarine, Soviet 92, 167

Yangel design bureau 72
Yankee Class submarine, Soviet 71, 74-5, 75, 77, 82, 84, 94, 95
Yom Kippur War 105-6, 148

Zulu Class submarine, Soviet 93-4, 95

Acknowledgments

The publishers would like to thank the following people who helped in the preparation of this book: Adrian Hodgkins, the designer and Ron Watson who compiled the index. The illustrations were supplied by the public relations departments of various manufacturers and by agencies of the US Department of Defense. The publishers would like to thank: Avco Corporation, Aerospatiale, Boeing, Bofors, British Aerospace, Avions Marcel Dassault, Ford Aerospace, General Dynamics, Goodyear Aerospace, Hughes Aircraft, Kawasaki, Lockheed, Martin Marietta, McDonnell Douglas, MBB, Rockwell, Shorts Brothers, Texas Instruments, Vought Corporation, Westland Helicopters. In addition thanks are owed to the following for the illustrations on the pages noted.

ECPA (via MARS): 128, 140 top
Peter Endsleigh Castle: 147 bottom
Deutsches Museum, Munich: (via MARS): 6-7, 12 top
Christopher Foss: 66-67, 88-89, 90, 103 top right, 108 bottom, 117 right, 142-143, 145 top, 154, 155 both
Ian V. Hogg: 9 both, 10 top, 125 bottom
Imperial War Museum, London: (via MARS): 12 lower
Military Archive Research Services, Lincs.: 1, 61, 63 all four, 100-101, 124, 125 top, 132, 138-139, 144 all three, 162 bottom, 176 both, 177 lower, 178-179
Swedish Air Force: 131 top
Swedish Army (via MARS): 141 bottom
Swiss Air Force (via MARS): 109 top left
Swiss Army: 140 bottom (MARS)
TASS: 69, 73, 83